Pain, Pleasure,
and the Greater Good

Pain, Pleasure, and the Greater Good

*From the Panopticon to the Skinner Box
and Beyond*

CATHY GERE

The University of Chicago Press
Chicago and London

The University of Chicago Press, Chicago 60637
The University of Chicago Press, Ltd., London
© 2017 by The University of Chicago

Published 2017
Printed in the United States of America

26 25 24 23 22 21 20 19 18 17   1 2 3 4 5

ISBN-13: 978-0-226-50185-7 (cloth)
ISBN-13: 978-0-226-50199-4 (e-book)
DOI: 10.7208/chicago/9780226501994.001.0001

Library of Congress Cataloging-in-Publication Data

Names: Gere, Cathy, 1964– author.
Title: Pain, pleasure, and the greater good : from the Panopticon to the Skinner
    box and beyond / Cathy Gere.
Description: Chicago ; London : The University of Chicago Press, 2017. | Includes
    bibliographical references and index.
Identifiers: LCCN 2017009044 | ISBN 9780226501857 (cloth : alk. paper) |
    ISBN 9780226501994 (e-book)
Subjects: LCSH: Utilitarianism—History. | Utilitarianism—England—History. |
    Utilitarianism—United States—History. | Psychology and philosophy. |
    Philosophy and science. | Medicine—Philosophy.
Classification: LCC B843 .G46 2017 | DDC 144/.6—dc23
LC record available at https://lccn.loc.gov/2017009044

♾ This paper meets the requirements of ANSI/NISO Z39.48-1992 (Permanence
of Paper).

For C.M.H.V.G.,
for her patience and forbearance

CONTENTS

INTRODUCTION

Diving into the Wreck

I came to see the damage that was done
And the treasures that prevail.
—Adrienne Rich, "Diving into the Wreck," 1973

THE TREASURES THAT PREVAIL

March 8, 1973, dawned gray and mild in Washington, DC, but on the political horizon conditions were ominous, with headlines roiling on the front page of the *Washington Post*. A story about POWs coming back from Vietnam was accompanied by a photo of an ecstatic young woman being swept off her feet by her war-hero husband, but there was no glossing over the fact that America had just suffered the first military defeat in its history. Three headshots running down the right side of the page represented a chain of people linking the Watergate break-in with the president's personal attorney, finally rendering Nixon's involvement in the affair an absolute certainty. Below that, in a piece about the dire economic outlook, the Federal Reserve chairman urged "fast action . . . to restore confidence in paper money." Meanwhile, in South Dakota, "the government sternly warned Indians holding the Wounded Knee settlement today that they must be out . . . or face the consequences."[1] From the

1

humiliating outcome in Vietnam, to the ballooning Watergate scandal, to the tanking economy, to the attention the American Indian Movement kept drawing to the genocide on which the United States had been founded, the nation was plunging fast into a full-blown crisis of moral legitimacy.[2]

A little before 9:30 a.m., Edward Kennedy arrived at a stately, wood-paneled room on the fourth floor of the Dirksen Senate Office Building, just northeast of the Capitol, and settled into his day's work. Kennedy, the Democratic senator from Massachusetts, was then forty years old, his life shattered by the assassinations of two of his older brothers, his back barely healed from being broken in a plane crash, his reputation shadowed by the fatal car accident on Chappaquiddick Island only four years before. His presidential ambitions scuppered by tragedy and scandal, he had recently embarked on the second act of his political career, focusing on medical reform through his chairmanship of the Senate Subcommittee on Health. In that capacity, he was grappling with an issue that seemed to exemplify the prevailing national mood of crisis, paralysis, and paranoia. The four-decade-long "Tuskegee Study of Untreated Syphilis in the Male Negro" had finally hit the headlines the previous summer, but despite official handwringing and public outrage, nothing had yet been done even to obtain *treatment* for the research subjects, let alone redress. In response to this foot-dragging, and hoping to clear up the matter once and for all, Kennedy had organized a series of investigative hearings on the topic of human experimentation.

March 8 was the sixth and last day of the hearings. Dozens of witnesses had already testified. A woman on welfare who had been forced onto the injectable birth-control drug Depo Provera told her story and described the debilitating side effects. Double-helix discoverer James Watson talked about the dangers and promise of genetic modification. The aristocratic British rabble-rouser Jessica Mitford delivered a scathing report on prison experimentation. Three ex-convicts testified about taking part in pharmaceutical trials in Philadelphia jails. The director of the National Institutes of Health discussed research funding. Various senators proposed schemes for legal oversight of scientific research. Now, finally, on the last day of the proceedings, two survivors of the Tuskegee Study were to bear witness to their experiences. Their lawyer—the African-American civil rights attorney Fred Gray—thanked Kennedy, noting for the record that this was the "first time that any governmental agency has permitted them to present their side of the story."

Sixty-seven-year-old Charles Pollard was the first Tuskegee survivor to speak. He described the start of the study, more than forty years earlier: "After they give us the blood tests, all up there in the community, they said we had bad blood. After then they started giving us the shots and give us the shots for a good long time." Pollard was unable to recall exactly the schedule of the shots, but one thing did stick in his memory: "After they got through giving us those shots, they gave me a spinal tap. That was along in 1933." Kennedy sympathized, noting that he had also undergone a spinal puncture—"they stick that long needle into your spine"—and knew that it was "rather unpleasant." Pollard admitted, "It was pretty bad with me," remembering that he had spent "10 days or two weeks" in bed after the procedure.

The second survivor to testify, Lester Scott, described how he had "taken shots" from the researchers every other week for decades, "figuring they were doing me good." Kennedy then asked: "You thought you were being treated?"

"I thought I was being treated then," Scott replied.

"And you were not."

"I was not."

"That is not right, is it?"

"No, it is not right."[3]

It is rare that the historical record presents us with such a clear and significant turning point. The syphilis study is now so notorious that the single word "Tuskegee" evokes a nightmare of the healing art gone horribly awry, of medical research conducted with contemptuous disregard for the health and well-being of the subjects, of a scientific establishment so profoundly racist that it was prepared to sacrifice hundreds of African-American lives on the altar of knowledge of dubious value. When the senator from Massachusetts and the farmer from Alabama agreed that what had happened was "not right," a line was crossed. In front of that line stand our current standards of scientific research ethics; behind it lies a shadowy Cold War America in which thousands of men, women, and children were pressed into service as unwitting laboratory animals.

As a result of the Kennedy hearings, the National Research Act was passed, mandating informed consent. From that time forward, scientists have been legally obligated to obtain the free, prior, voluntary, and comprehending agreement of their research subjects. This was a change in day-to-day ethical practice of lasting significance. Mandatory informed consent has spread from research to therapy, from life science to social science. It underpins definitions of sexual harassment and rape; it has emerged as the preferred structure for negotiations between nation-states and indigenous peoples; it may provide the most intuitive framework for practical ethical action in the contemporary world.[4] Tempting as it is to be cynical about the routine, bureaucratic, and often excessive practice of informed consent today, it is also difficult to imagine life without it. On March 8, 1973, a flawed, death-haunted man swam down to the moral nadir of liberal democracy, and salvaged mandatory informed consent from the wreckage.

MEDICAL UTILITY

Despite, or perhaps because of, all that has been said and written about informed consent, a host of confusions, obfuscations, exaggerations, and denials still spin around that moment of agreement between Lester Scott and Edward Kennedy. This book will argue that without a fresh understanding of what was at stake in that encounter, we are in danger of letting its most important lessons slip out of our grasp. The key to the analysis is a simple insight: the drama of stopping Cold War human experimentation was not a battle between good and evil, but rather a conflict between two conceptions of the good. Both were philosophically defensible. One was victorious. And when the other was defeated, it was consigned to the dustbin of incomprehension. In a breathless sprint for the moral high ground, scholars have denounced the Tuskegee researchers as incorrigible racists, unable or reluctant to understand their actions in any other terms. This has made it harder to analyze with any precision what went wrong beyond the question of racial prejudice. And now, as a result, the core problem that informed consent was supposed to solve is making a stealthy but steady comeback.

Tuskegee is such a loaded subject that it is well-nigh impossible to present a neutral statement of the facts of the case. In 1932, after a syphilis treatment program in the community was derailed by the Great

Depression, government doctors identified around four hundred black men with latent-stage syphilis, presumed to be noninfectious, in Macon County, Alabama, and embarked on a study of the progression of the untreated disease, using two hundred uninfected men as a control group. At the time, the best treatments for syphilis (mercury rubs and derivatives of arsenic) were dangerous, burdensome, and unreliable. Told only that they had "bad blood"—the local vernacular for syphilis— the subjects were given aspirin and iron tonics, along with some deliberately inadequate treatment, under the guise of proper medical care. One purpose was to see if late-stage syphilis would, as many physicians assumed, spontaneously remit without treatment, and if so, in what percentage of cases. Another goal was to ascertain racial differences: a study of untreated syphilis had already been done in Norway; the Tuskegee researchers wanted to compare those results in whites with the progression of the disease in blacks. More specifically, the consensus was that whites suffered from neurological complications from late-stage syphilis, while blacks were more likely to experience cardiovascular problems.

At the end of the first year, the subjects were sent a letter telling them to meet the nurse at a designated spot so that she could drive them to Tuskegee Hospital for an overnight visit. "REMEMBER," the letter announced, "THIS IS YOUR LAST CHANCE FOR SPECIAL FREE TREATMENT." The "treatment" was, in fact, the administration of a diagnostic spinal tap, looking for evidence of neurosyphilis. This is an extremely unpleasant procedure, and in 1933 it was cruder than today. Side effects included leakage of cerebrospinal fluid, causing headaches, nausea, and vomiting that could last a week or more. The spinal taps should have heralded the end of the study, but the man who had confected the deceptive letter persuaded his bosses that "the proper procedure is the continuance of the observation of the Negro men used in the study with the idea of eventually bringing them to autopsy."[5] In the end, the research went on for decades, long past the time that penicillin became available as an easy-to-administer and effective treatment.

There is no evidence that anyone in the study was deliberately infected with the disease; there is abundant evidence that strenuous measures were taken to keep the men away from antibiotics. In 1932 it was thought that latent syphilis would remit in a percentage of long-term cases; by 1953 it was acknowledged that men in the infected group were

Figure 1. Fred Gray, attorney for the Tuskegee Study survivors, with Martin Luther King Jr., around 1955. Gray defended Rosa Parks and worked with King on the legal dimensions of the civil rights agenda. His decisive role in bringing the Tuskegee Study to a close connects informed consent to the struggle for voting rights. Medical ethics reform was part of the broad postwar challenge for the United States to live up to the universalistic philosophy on which it had supposedly been founded. Reprinted from *Bus Ride to Justice* by Fred Gray, with the permission of NewSouth Books.

dying faster than the controls. Still, nothing was done to treat them. Finally, in 1972, a whistle-blower at the Public Health Service informed a journalist friend, whose news agency wrote a story about the research that was published on the front page of the *New York Times*. Coming on top of a host of prior research scandals, and with its explosive racial dimension, the Tuskegee exposé ensured that informed consent in medical research became the law of the land.[6]

In 1978 the crusading historian Allan Brandt rendered his verdict on the cause and meaning of these events: "There can be little doubt that the Tuskegee researchers regarded their subjects as less than human. As a result, the ethical canons of experimenting on human subjects were completely disregarded."[7] Most subsequent retrospective accounts of the study have followed Brandt's lead, arguing that racist white doctors dehumanized black people and were therefore comfortable with violating the usual standards of research ethics.

Without denying the importance and timeliness of Brandt's denunciation of scientific racism, I venture to suggest that his analysis of the Tuskegee researchers' motives is a little sweeping. One aspect of the story that has vexed and puzzled commentators since the time of the exposé, for example, is the enthusiastic involvement of African-American medical professionals in the study. Surely nurse Eunice Rivers, the much-adored liaison between doctors and research subjects, who lived through the controversy and remained staunch in her defense of the research, did not simply regard the participants as "subhuman"? Nor, I am convinced, did the widely respected doctor Eugene Dibble, who managed the project at the Tuskegee Institute Hospital. Dibble died of cancer in 1968, four years before the exposé, after a long career devoted to improving the health of his community.

Historian Susan Reverby has a chapter about Dibble in her valuable 2009 book *Examining Tuskegee*. In it, she asks this pivotal question about the doctor's motives: "Is it possible that Dibble was enough of a science man that he was willing to sacrifice men to the research needs, justifying the greater good that would hopefully come from the research?"[8] Notwithstanding the slightly incredulous tone in which the question is posed—"is it possible?"—the answer must be a straightforward "yes." Dibble regarded himself as working for the greater good. He believed that scientists and doctors could and should make decisions on behalf of others in order to maximize societal benefit. Dibble was, in other words, a *utilitarian* when it came to the ethics of medical research on human subjects. Utilitarianism is the moral creed that considers, above all, the consequences of any given course of action for the balance of pain and pleasure in the world. In contrast to the idea of absolute rules of conduct—that everyone has a right to informed consent, for example—utilitarianism asks us to weigh the costs and benefits of our actions and choose the path that will maximize happiness and pleasure or minimize pain and suffering.

To understand the ideology that conscripted six hundred black men into the medical research enterprise for forty years without their knowledge or consent, we need to locate Tuskegee in the larger context of medical utilitarianism. Before the Kennedy hearings, the utilitarian credo "the greatest good for the greatest number" was *the* driver of most human experimentation in the United States. The cost-benefit analysis that is the essence of utilitarian morality found its most compelling application in scientific medicine. In the face of the devastation wrought by uncontrolled disease, a medical breakthrough was the clearest example of an outcome that justified whatever it took to get there: the harm to research subjects was limited; the payoff was potentially infinite. Not only does this ethos help explain the participation of African-American doctors and nurses in the study, it also accounts for the thousands of Americans of every race drafted into similar research projects around the mid-twentieth century. Eugene Dibble was not a race traitor, psychological puzzle, or victim of false consciousness. He was a good physician working hard for the betterment of his people within the widely accepted (if not explicitly articulated) framework of medical utilitarianism. As one distinguished pediatrician recently recalled, reflecting on research he undertook with children in an orphanage in the early 1950s, "I use the hackneyed expression that the end justifies the means. And that was, I think, the major ethical principle that we were operating under."[9]

Since 1973 a small number of articles, mostly in medical journals, have, in fact, defended the Tuskegee research in precisely these terms. Summarizing these in 2004, University of Chicago anthropologist Richard Shweder asks, "What if, consistent with the moral logic that every person's welfare should be weighted equally, the PHS researchers would have made exactly the same utilitarian assessment, regardless who was in the study [black or white]?"[10] Shweder seems to think that an affirmative answer to this question would amount to a full exoneration. If the scientists would have made "exactly the same utilitarian assessment" regardless of the racial identity of the subjects, then no wrong was done.

Even if we accept Shweder's version of the story, I think we can still agree that something went morally wrong. He is correct that the Tuskegee subjects were not drafted into the syphilis study *just* because they were African-American. They faced this predicament as a result of a systematic bias built into utilitarian medicine that permitted *any* vulnerable citizen to be sacrificed on the altar of science. Time and again, nonconsensual research on terminally ill, marginal, impoverished, incapacitated, under-educated, or institutionalized human subjects was rationalized on cost-benefit grounds: terminal patients would die soon anyway; poor rural populations were otherwise out of reach of medical care; illiterate people could not be expected to understand risk; state asylums were already rife with communicable diseases; the lives of institutionalized people with disabilities were so unpleasant that research procedures represented little extra burden. Even when such assessments were done with care and sensitivity, even when the harms and benefits were weighed with sincerity, gravity, and good will, under the terms of this calculus, anyone with little enough to lose was fair game in the cause of medical progress. The legacies of slavery in Macon County, Alabama—poverty, illiteracy, medical neglect—were a natural fit with this larger system of moral justification.

During the buildup to Tuskegee, as scandal after medical scandal broke in the press, utilitarianism *was* often identified as the core ethical problem in human experimentation. But in the early 1970s, when all the issues of scientific power and responsibility were finally put on the table, utilitarianism was conflated with fascism, and the opportunity to critique it on its own terms was lost. Leading up to that moment of reckoning between Edward Kennedy and Lester Scott were the aftermath of World War II, the rise of the civil rights movement, and the dissolution of the European empires. The philosophically complex question of the rights and wrongs of medical utilitarianism got lost in the gravity and urgency of the issues—racism, Nazi atrocities, civil rights—that were invoked in agonized discussions of the Tuskegee Study.

There is no denying the historical freight carried by the human experimentation controversy. The idea that informed consent should be an absolute mandate originated at the 1947 Nuremberg Medical Trial, at which twenty-three Nazi physicians and medical administrators were prosecuted by an American legal team. At the end of the proceedings, one of the judges read out a ten-point code of research ethics, which had been drafted to serve as the foundation for the convictions. The first provision of the so-called Nuremberg Code was an emphatic command: "The voluntary consent of the human subject is absolutely essential."

Perhaps forgivably, the medical profession on the side that had defeated fascism did not regard the Nuremberg Code as binding, however. After the trial, most American research scientists proceeded to ignore the requirement to obtain consent from their human subjects. Instead, they held to the prevailing norms of their scientific vocation, according to which they were trusted to judge the balance of harm to subjects against good to future patients.

For a decade or so, this rather flagrant bit of victors' justice went unremarked. But as civil rights cases began to accumulate in the courts, domestic violations of the right to informed consent could no longer be ignored. Starting in 1963, human experimentation scandals began to break with disconcerting frequency. Radicalized lawyers denounced medical researchers as human rights abusers. Journalists drew invidious comparisons between Nazi medicine and American research. When news of the Tuskegee Study broke in 1972, it seemed to confirm the diagnosis of fascist callousness. If American scientists could treat the descendants of slaves with such lethal contempt, how different could they be from their Nazi counterparts? The Kennedy hearings on human experimentation followed. The first draft of the legislation mandating informed consent took its wording straight out of the Nuremberg Code.

It took outraged comparisons with the Nazis to stimulate the root and branch reform of medical research ethics; nothing less could overcome the professional autonomy of American medicine. But once the necessity for reform had been accepted, both sides backed off. Comparisons with the Nazis were wide enough of the mark that the activists gained no legal traction, while the medical profession quietly embraced informed consent. In the quiet after the storm, utilitarianism was left momentarily chastened but basically intact. As it turned out, the calculation of benefits and harms proved to be so deeply woven into scientific medicine, and so necessary to its operation, that it was impossible to disentangle it neatly from the fabric of medical practice and hold it to the light.

A peace treaty was duly organized. In 1974 a commission was appointed to hammer out a philosophical foundation for the new medical ethics. Four years later, the group issued its report. Medical ethics was founded on three timeless and universal principles, it announced: autonomy, beneficence, and justice. "Autonomy" meant that individuals have the right to decide what happens to them. Securing informed consent was therefore the primary duty of all investigators. "Beneficence" dealt with the risks and rewards of the research, and demanded that any given study satisfy a cost-benefit analysis. Here was the guiding principle of medical utilitarianism, now demoted to second place in the ethical ranking. Finally, "justice" required that research subjects not be drawn exclusively from the most vulnerable members of society. As a matter of practical action, this third principle was quickly assimilated into the procedural niceties of informed consent. Anyone whose freedom was compromised by exceptionally difficult life circumstances was (at least in theory) protected from exploitation under guidelines barring "undue inducement"—for example, monetary payment to people in great financial need or commutation of sentences for prisoners.

With the principle of justice absorbed into informed consent, the ultimate importance of the commission's report lay in the fact that it ranked autonomy above beneficence. By placing autonomy first—as the Nuremberg Code had done—it effectively made informed consent a "threshold condition" for any sort of human experimentation.[11]

Since then, the preeminence of informed consent has been resoundingly confirmed. The ranking of the principles is now firmly established. Autonomy rules in contemporary research ethics and clinical practice, but utilitarianism, in the guise of "beneficence," happily coexists with it as a close runner-up. Meanwhile, unvarnished utility calculations still underpin many aspects of modern medicine, from medical resource allocation and emergency medicine, to responses to epidemics, global health initiatives, and most domestic public health measures.

Making peace with utilitarianism was thus a necessary compromise with the bedrock realities of modern medicine. As a piece of pragmatic moral reasoning, the so-called principles approach, with informed consent at its heart, works extremely well. The ever-expanding regulatory reach of informed consent may be the most important development in practical ethics in our secular age. But I think it is important to understand not only that it is right, but also *why*. The demotion of utilitarianism from first to second place in the ethical hierarchy of modern medicine offers us a golden opportunity to understand how, and to what extent, greater-good reasoning fails as a guide to action. We can theorize about such matters, but the history of medicine offers us a wealth of data about how such moral abstractions play out in real life.

Utilitarian reasoning will never be exorcised from modern medical research. By its very nature, a clinical trial weighs risk and benefits to present subjects against potential benefit to future patients. Cost-benefit calculations cannot be avoided in the context of modern medical provision, from rationing drugs in socialized medical systems to setting insurance premiums in privatized ones. Utilitarianism and scientific medicine are deeply intertwined, interdependent, and to a great extent inseparable. The delicate trick is to balance acknowledgment of utilitarianism's indispensability with an understanding of why informed consent arose as a necessary corrective.

I find it helpful to think of the relation between informed consent and utility as analogous to that between figure and ground: in the ubiquitous diagram, two silhouetted profiles face one another; the space between them has the shape of a vase. Which do you see? Vase or faces? Faces or vase? Imagine that the profiles are those of doctor and patient, negotiating a treatment. The image of two human faces, engaged in an exchange as equals, represents the practice of informed consent. Utility is the vase, the course of action that emerges once the two sets of concerns and interests are brought into symmetry, made mirror images of each other. Informed consent says that doctor and patient must engage in truthful dialog until they view the situation in the same way, and have agreed on the path ahead. Then, and only then, does the vase come into view.

At his arrogant worst, the utilitarian focuses exclusively on the vase. Assuming that he knows the contours of a rational profile—mirroring his own, of course—he believes the vase's dimensions can be calculated with just pencil and paper, without the need to negotiate with a human being of equal worth. It is the purpose of this book to persuade you that this is a moral error as devastating as it is subtle.

JEREMY BENTHAM

I first became interested in the history of medical utilitarianism during a postdoctoral fellowship in medical ethics and health policy at Cambridge

University. Our research group was formed in response to an ongoing scandal in the British medical profession. In the course of an inquiry into poor outcomes in a pediatric cardiology unit, it emerged that certain British hospitals, under the auspices of the National Health Service, had routinely removed and stored the internal organs of children who died within their walls, without informing relatives or asking their permission. Bereaved parents began to call the hospitals where their children had been treated. It turned out that one particular hospital in Liverpool had retained almost all of the internal organs of children who died there. In December 1999 the coroner for Liverpool suggested that the practice was unlawful. There was a media outcry, the trickle of phone calls from parents turned into a flood, and the government announced that there would be an independent inquiry into organ and tissue retention in the NHS.

A support group for affected parents was set up, which obtained a change in the law to allow for the cremation of organs after the burial of a body. Many of the parents subsequently went through second, third, and even fourth funerals as parts of their children's bodies—from organs and organ systems down to tissue blocks and microscope slides—were returned, literally piecemeal, by various hospitals around the country. As a result of these events, and the sympathetic government response, the informed consent revolution that had swept the American medical profession in the 1970s finally made its way into every aspect of medical practice in Britain.

In 2001 the Wellcome Trust, a British medical charity, put out a call for proposals to investigate aspects of the crisis with a view to contributing to the discussion about policy reform. I was coming down the home stretch of a PhD in the history and philosophy of science at Cambridge, and was interested to see if my hard-won skills had any real-world application. A colleague and I applied for and received a grant to explore the legal and ethical dimensions of human tissue storage, using the brain bank at the Cambridge University Hospital as a case study.

Our first step was to try to understand the historical preconditions and precedents for the scandal. It turned out that the legislation governing tissue collection practices had its roots in the Anatomy Act of 1832, a piece of legal reform inspired by Jeremy Bentham, the most famous and influential exponent of utilitarian philosophy, sometimes known as the father of utilitarianism.[12] This direct connection to the history of utilitarianism made the situation unfolding before our eyes—with utility and autonomy so clearly opposed—seem especially stark.

That the organ retention scandal was about the bodies of the dead rather than the living also sharpened the dichotomy between cost-benefit thinking and the enshrining of inalienable rights. The use of archived human tissue for research purposes has such a strong utilitarian justification that an explicit appeal must be made to another, overriding principle, such as the right of citizens in a liberal polity to dispose of their remains or those of their kin as they see fit. As a consequence of this involvement with a homegrown medical scandal and its fallout, utilitarianism became visible to me not just as a philosophical abstraction but as a political reality, with very material consequences for the practice of scientific medicine.

Primed by this close encounter with the implementation of informed consent in Britain, I continued to register a tension between utility and autonomy in medical ethics after I moved to the United States. My first academic teaching job was at the University of Chicago, where I attended the weekly meetings of the ethics consulting team at the University Hospital. One doctor repeatedly registered his protest at the terms of reformist medical ethics. Was it really the place of the medical profession to promote autonomy above all other values? Surely it should dedicate itself first and foremost to the relief of suffering? Another member of the group was an eloquent champion of patient autonomy, but even she returned again and again to the same question: was not anyone who refused the hospital's beneficent treatment irrational by definition, and therefore unable to give informed consent? Despite a deep institutional and professional commitment to it on the part of the institution, patient autonomy seemed a fragile thing amid the clamoring reality of a large urban hospital.

As I listened to the deliberations of the ethics team at the University of Chicago Hospital, it struck me that the issues dividing the assembled doctors were as much about human nature as about moral principle. Underlying each position voiced around that conference table was an implicit set of views about human motivation, behavior, and decision-making. One weary veteran of such meetings has described these views as "competing and incompatible conceptions of the self." She observes that they "tend to be implicit and therefore largely invisible," but they tell us "what is important and what is not, what a human being is and how creatures of this sort should behave, and how we should respond when we are threatened with infirmity, illness and death."[13]

By inviting the Tuskegee survivors to Washington, and listening carefully to their testimony, Edward Kennedy enacted the view of human nature underlying informed consent. This perspective acknowledges that we are all suffering animals, crying out at the pain of a spinal puncture, and at the same time free beings with a right to self-determination. Kennedy's legislative activism in the cause of informed consent was completely consistent with the way he listened to the research subjects and gave them a voice in the democratic process.

Admirable though the senator's stance undoubtedly was, it has to be admitted that the stately chambers of Congress are a world away from the controlled chaos of a hospital emergency room. Sitting round the table with the Chicago doctors, I learned how delicate moral freedom is in a modern hospital setting. In truth, patient self-determination aligns poorly with scientific medicine even under the best of circumstances. Because the conditions of illness—pain, debility, delirium, nausea, and dementia—tend to set patients' decision-making abilities at their lowest ebb, medical paternalism is less occupational temptation than unavoidable burden. As the guardians and gatekeepers of medicine's rapidly increasing capacities, physicians are in an inherently authoritative position in relation to those they treat. At a more abstract level, the medical sciences owe much of their success to a mechanistic understanding of our biology, including how our bodies and brains produce our desires and decisions. This scientific focus on tight chains of cause and effect has no place for genuine freedom of action. So much for the "autonomous self" that should undergird the practice of informed consent.

The "utilitarian self," by contrast, is a perfect fit with medical science. Here is a famous passage from Bentham pondering, in 1789, the relation between human nature and ethics:

Nature has placed mankind under the governance of two sovereign masters, *pain* and *pleasure*. It is for them alone to point out what we ought to do, as well as to determine what we shall do. On the one hand the standard of right and wrong, on the other the chain of causes and effects, are

fastened to their throne. They govern us in all we do, in all we say, in all
we think: every effort we can make to throw off our subjection, will serve
but to demonstrate and confirm it. In words a man may pretend to abjure
their empire: but in reality he will remain subject to it all the while.[14]

How neatly this outlook dovetails with the goals, aspirations, and realities
of modern medicine! Where is the sovereignty of pain and pleasure more
obvious than in a hospital? When could our "subjection" to the "empire"
of these primordial sensations ever be more powerful than when we are
really sick? Surely the profession of medicine is the one arena when pain
and pleasure alone "point out what we ought to do." How could there
be a better fit between a theory of human motivation (pain-avoidance
and pleasure-seeking), a set of practical actions (treating the sick and
alleviating their suffering), and a system of ethical reasoning (the greatest happiness for the greatest number)? In contrast to the fragile and
often paradoxical notion of "patient autonomy," medical utilitarianism
is *seamless*.

THE SCIENCE OF PAIN AND PLEASURE

At the same time as I was attending ethics committee meetings at the
University of Chicago Hospital, I was teaching two classes at the university, one about the history of medical ethics and the other about the history of neuroscience. The two sets of readings were mostly very different,
but at a series of intriguing points they seemed to overlap. Given that the
debates at the hospital concerned the constitution of human nature, I
was not surprised that neuroscience and psychology displayed an obvious moral dimension. A bit more surprising was how often utilitarianism
seemed to be at stake. The transcripts of the Kennedy hearings contained
a particularly striking example of this tendency of medical ethics and
neuroscience to find their common ground in debates about the greater
good.

On the third day of the hearings, a couple of weeks before Kennedy's exchange with the Tuskegee survivors, the focus of the proceedings
turned to research on the human brain and nervous system. Some of the
most distinguished neurologists, neuroscientists, and psychologists in
America were summoned to testify. Of all the scientists who spoke at

the hearings, these men alone gave a resounding defense of utilitarian research. A decade of experimentation scandals had taught the medical profession that greater-good arguments were not going to fly. But somehow neuroscientists had not received the memo. They seemed unable even to imagine justifying their work on any other grounds.

Two of the witnesses were particularly explicit about the scientific reasoning behind their devout utilitarianism. In the morning session, Robert Heath, chair of psychiatry and neurology at Tulane University, gave a presentation on his research. Heath explained that his laboratory had identified the pain and pleasure pathways of the brain, and that electrically stimulating these areas allowed him to control the happiness and misery of his patients. He described how he had implanted electrodes deep in the brains of psychiatric patients at the Charity Hospital in New Orleans and left them there for months and sometimes years for the purpose of manipulating mood and behavior. To demonstrate the power of his technique, he showed three short films of the experiments, in which the patient-subjects, electrodes buried in heavy bandages around their heads, seemed to transition abruptly from apathy to rage, from misery to hilarity, as the current was switched on and off.[15]

Declaring himself a "healer" above all, Heath explained that he had implanted the electrodes under the moral imperative to pursue the greater good, experimenting on these patients for the soon-to-be-realized benefit of such people everywhere. Not only were his experiments morally permissible, he argued; it would be unethical of him *not* to do them, and thus to deprive humanity of the fruits of his work. Moreover, these particular patients were uniquely qualified to serve as experimental subjects. Because "life in the back ward of a chronic mental hospital is worse than death," Heath claimed, any experimental therapy, even a clearly dangerous one, "is worthy of trial."[16] For this neurologist, the end always justified the means. As a healer and a scientist, working with a patient population whose conditions of existence he judged to be worse than death, he was supremely qualified to weigh the balance of harms to his subjects versus benefits to the whole human race. His version of medical utilitarianism was as pure and as extreme as was possible.

After Heath had explained and justified his work, the famous Harvard psychologist B. F. Skinner gave his testimony. The control of human behavior through ordinary reward and punishment was just a fact of life, he declared. He assured Kennedy that the current worries about social control were therefore founded on a major philosophical error. "We are all controlled, all the time," he flatly asserted. The idea of psychological autonomy was just an illusion, and a dangerous one at that. With ultimate consequences firmly in mind, the job of the behavioral scientist was to accept the ubiquity of psychological control, and use his insights into the power of reward and punishment to design a better world.[17] That better world was defined in classically utilitarian terms: the only measure of progress was the balance of pleasure and pain. Between them, Heath and Skinner laid out a completely Benthamite view of human nature and purpose.

Kennedy was politely receptive to the two scientists, but outside the Senate chambers Heath and Skinner found themselves running the gauntlet of public outrage. Civil rights activists hounded Skinner's public appearances. Harvard had to offer him police protection. He was burned in effigy. He was denounced as a fascist in the *New York Review of Books*. Heath fared even worse. After the hearings, a muckraking local journalist in New Orleans published a long article describing how Heath's "lackeys" scoured the wards of the hospital for fresh "victims."[18] Tulane students began to demonstrate against his presence on campus. In 1977 one of his colleagues gave a paper at a world congress claiming that Heath had violated "all standards of ethics."[19] He hung on at the university, but by the time he retired in 1980 he was professionally isolated and could not find a publisher for a manuscript explaining his achievements and defining his legacy.[20] Now, just over forty years later, the films that Heath proudly played for Congress are considered so controversial that Tulane's lawyers responded to my request to view them by locking down the whole collection for seventy years on the grounds of "patient privacy."[21]

Of all the researchers called upon to testify at the Kennedy hearings, it was only the psychologists who were prepared to defend a pure version of medical utilitarianism. Without exception, they defended their work on the grounds that the greater good it would bring to society should prevail over concerns for the research subjects' rights and freedoms. Heath and Skinner's testimony provides the clue as to why. Their views about right and wrong drew not from philosophical ethics but from evolutionary biology, behavioral psychology, and the neuroscience of reward pathways. Utilitarianism was merely the logical outgrowth of their understanding of human behavior as driven by pain-avoidance and pleasure-seeking. Theirs was a complete system of moral, psychological, and biological justification—a set of values so deeply embedded in the natural world that it looked to them—and, indeed, to Senator Kennedy—like nothing more or less than value-free science.

As I pored over the Kennedy hearing transcripts, I began to wonder if Heath and Skinner could be connected to the longer history of utilitarian medicine that began formally with the 1832 Anatomy Act. It seemed a long shot, but there was something so striking about the way they had turned the ethos of the greater good into a set of facts about human nature. The key was handed to me—as so often happens—by a graduate student. In the history of neuroscience class, we were reading about nineteenth-century Britain. I asked if anyone knew anything about Alexander Bain, the so-called father of scientific psychology, and one of the students offhandedly remarked that she didn't know anything about him "except that he was a utilitarian."[22]

Bain, it turned out, unlocked the whole puzzle. A devout follower of Bentham, he transposed all of the master's ideas about ethics and politics into the language of scientific psychology. In the 1850s, in a series of highly influential textbooks, he hammered out a complete utilitarian view of human nature, based in the physiology of pain and pleasure. And he was not alone. A handful of thinkers came before him; a legion of experimenters followed. In 1898 one of these experimenters went on a lecture tour of the United States. A graduate student named Edward Thorndike attended the lecture and decided to take up the same research program on a larger scale. Thorndike appropriated utilitarian psychology, claimed it as a novel invention, and gave it a distinctively American twist. Skinner followed in his footsteps at the same university. And so, step by step, from Bentham to Bain to Thorndike to Skinner, British utilitarianism evolved into American behaviorism, in a line of direct intellectual descent. Historically speaking, utilitarianism is not just an arid system of moral reasoning but a multidimensional—and in many ways persuasive—account of the driving motivational force of Bentham's "two sovereign masters, *pain* and *pleasure.*"

My interest in the history of utilitarian ethics in medicine had now brought me face to face with something even more deeply embedded in scientific practice. Utilitarianism did not just sit, so to speak, "on top" of medical ethics, as a guide to action that had little to do with the actual object of inquiry. Rather, it arose out of a scientific view of the fundamental psychological makeup of the human animal. By the twentieth century, pain and pleasure were understood as the most basic of biologically programmed drives. Moreover, there was no escaping the forces of reward and punishment; the idea of psychological or moral freedom was nothing but a prescientific superstition. Within this scheme, greater-good reasoning appeared as a fact of nature, so mixed up with the experimental protocols used to investigate pain and pleasure that it was rendered invisible. It is this multidimensional utilitarianism—a moral, psychological, physiological, and biological theory tying together pain, pleasure, and utility—that constitutes the subject matter of this book.

As a frame and foundation for the whole argument, the first chapter makes medical utilitarianism visible through the story of its opposite—the doctrine of informed consent. It tells of a series of legal and regulatory showdowns, beginning with the Nuremberg Medical Trial. Following the trajectory of American medical research ethics after the trial, the chapter details a handful of the many experiments conducted in the 1950s without benefit of informed consent. It then examines how things began to change in the early 1960s, when the rise of civil rights consciousness made American violations of the Nuremberg Code impossible to ignore. Finally, it returns to where this introduction began, to the 1973 Kennedy hearings and the decisive victory of patient autonomy over medical utility.

The chapter ends with the testimony of the brain scientists at the Kennedy hearings, probing the psychological and biological assumptions that underlay their extreme utilitarian ethics. Forced to articulate a defense of their access to experimental subjects, these men sketched out a highly coherent intellectual system, with its foundation in the physiology of pain and pleasure and its capstone in extreme utilitarian justifications of their work. The rest of the book tells the story of the long life, ignominious death, and recent resurrection of this worldview.

PAIN, PLEASURE, AND UTILITY

The first chapter ends in 1973 with the demise of utilitarian medicine and the triumph of patients' rights. Chapter 2 travels back in time to the origins of the conflict between utility and rights, on another continent, two centuries earlier. The year that the French Revolution broke out, Jeremy Bentham published his famous treatise proposing the test of utility as the basis for rational social reform. As French politics subsequently swung violently leftward, Bentham drew more emphatic attention to the contrast between practical utility and revolutionary rights. The idea of a human right, he declared, was an empty formula that led straight to the kind of confusion and anarchy that was convulsing France. In contrast to the metaphysical confusion of rights talk, the experience of pleasure and pain was real, palpable, and measurable. Utility was the only rational test of progress. It could be quantified, and then applied directly to questions of politics and law.

Bentham often expressed his utilitarian convictions in medical terms. He talked of "moral physiology" and "moral pathology," and he sketched the idea of a "moral thermometer" that would measure degrees of delight and misery. Convinced that utilitarianism could work through manipulation of individuals, he also started to think through how reward and punishment shape behavior. In 1798 he developed his most audacious antidote to revolution—an elaborate proposal for a vast grid of pauper workhouses in which pleasure and pain would be deployed to induce the poor to discipline *themselves*. With Britain at war with France and facing the imminent threat of proletarian insurrection at home, Bentham's utilitarianism presented itself as a scientific prophylactic against this double menace.

Chapter 2 ends with the victory of British utilitarianism under the shadow of the guillotine; chapter 3 explores how the counterrevolution was fought and won over the body of the industrial pauper. In 1798 Bentham was joined by a fellow utilitarian, the Reverend Thomas Malthus, who presented an even gloomier assessment of the hazards of rights-based politics in his *Essay on the Principle of Population*. In their different ways, Malthus and Bentham argued that the only way to deal with poverty, misery, and vice was the rigorous application of the test of utility. Between them, they developed the biology, psychology, and ethics that would define the utilitarian self right up to the 1970s. One of Bentham's most substantive legal victories was the aforementioned 1832 Anatomy Act, which made the bodies of people too poor to pay for their own funerals available to medical schools for dissection. The chapter explores popular resistance to the act, underscoring the way that the utilitarian calculus singles out the poorest and most vulnerable members of society for sacrifice on the grounds that they are the ones with the least to lose. It concludes with the triumph of the Malthusian-Benthamite workhouse system in the reforms of 1834.

Chapter 4 follows the story of the next generation of Bentham's and Malthus's disciples, including John Stuart Mill and Alexander Bain. It then details how utilitarianism was transformed into a full-blown experimental science in a laboratory on the grounds of the West Riding Lunatic Asylum in Yorkshire, where David Ferrier, one of Bain's protégés, electrically stimulated the brains of scores of experimental animals and theorized the motivational force of reward and punishment. By the end of the nineteenth century, a network of British scientists, economists, and philosophers had agreed on a vision of the nervous system as a stimulus-and-response mechanism, directed by a pain-pleasure system that had evolved to guide the organism toward survival and reproduction.

The penultimate chapter follows utilitarian psychology across the Atlantic, showing how American researchers developed it into a systematic and replicable experimental program. First the Harvard psychologist Edward Thorndike formulated his "Law of Effect," linking animal and human learning to reward and punishment. Thorndike's successor was B. F. Skinner, who transformed pain-pleasure psychology into a stunningly successful research agenda. In 1953 two researchers in Montreal discovered the "reward center" of the rat brain, opening up the possibility of direct electrical stimulation of neural circuits of pain and pleasure. Robert Heath quickly applied the technique to indigent psychiatric patients in the New Orleans Charity Hospital. By the mid-1960s, Heath had developed utilitarian psychology into its most extreme expression.

The story of the rise of utilitarian psychology in America is intertwined with that of an increasingly militant countermovement, marching in lockstep with the battle for informed consent in medical ethics. In the aftermath of the Holocaust, an influential group of social scientists became preoccupied with defining, understanding, and cultivating the qualities of "the democratic type," the ideal citizen of a liberal society. Preeminent among these qualities was psychological autonomy, the capacity for independent judgment. Behaviorism seemed to deny the very possibility of independent thinking, and throughout the 1950s and 1960s its proponents were increasingly denounced as latent fascists. The chapter ends with an episode in which all the strands of the controversy converged. In 1973 Heath published an article claiming that he had cured a young man's homosexuality by means of implanted electrodes in his reward pathways. His timing could not have been worse. The duration of the experiment exactly coincided with the campaign to remove the diagnosis of homosexuality from the "Bible of psychiatry." No prizes for guessing which side won.

When I embarked on this book, I assumed that it would end with the triumph of autonomy and defeat of utility exemplified by Heath's aborted sexuality research. But as I worked my Benthamite seam, I became aware that utilitarian psychology was undergoing a powerful renaissance.[23] Starting in the 1980s, and gathering irresistible force in the last couple of decades, a movement has emerged that unites economics, evolutionary theory, and ethics into a new science of consumer behavior. Currently enjoying an ever-widening influence over government policies in countries all over the world, this new field faithfully recapitulates all the discourse about biology, psychology, and utility in late-nineteenth-century Britain.

Accordingly, the final chapter explores the rebirth of utilitarian psychology in "neuroeconomics" and "behavioral economics." This new crop of utilitarian psychologists presents a very different kind of target from the Benthamites of Cold War medicine. But the fidelity with which they have returned to nineteenth-century British utilitarianism, combined with their outsize influence, renders the lessons of this history freshly urgent. I end the book with a historical critique aimed at these twenty-first-century revivalists, with the aim of making it just slightly harder for Bentham's two sovereign masters to reassert their rule.

THE ARCHANGEL AND THE PROLE

Before embarking on the story of the utilitarian self, I should disclose that this topic taps into longer-standing and more personal preoccupations than just postdoctoral research. Thirty-odd years ago, a few months into my undergraduate career at Oxford University, I read the philosopher R. M. Hare's *Moral Thinking: Its Levels, Method and Point*, and experienced an instantaneous conversion to his brand of utilitarianism. I read it lying in the grass of Christ Church Meadows, over the course of a sunny afternoon, transfixed by its serene conviction that ethics were as much a matter of logic and the facts as any area of human inquiry.

Wandering back to my rooms afterward, I bumped into a friend, and I have a vague memory of actually grasping his lapel in my fist before fixing him with a fanatical stare and saying, "Did you know that there is a *logic* to the word 'good'?" Captivated by Hare's rationalism, I became a zealot of utility and eventually dropped out of Oxford in order to devote myself to a militant political movement based on the standard revolutionary premise that the present must be sacrificed to the future, the few to the many, the rich to the poor, and so on. Luckily for me, an Oxbridge dropout earnestly pursuing this agenda in the yuppie 1980s was more comic than tragic, and I outgrew it before I had committed any atrocities in its name.

Of course, utilitarianism is as flexible as any other ethical system and need not lead down the rabbit hole of idiotic vanguardism. My choices at the time had at least as much to do with my youth and the company I kept as with the academic philosophy I was exposed to. But I think it is fair to say that I experienced firsthand the corrupting potential of the means-end relationship in utilitarianism. Under the terms of my crude calculus, all people, including myself, were but means to the end of a better future. Over the years of my belated maturation, I managed to grope toward some kind of alternative, based on respect for others and self-respect. When I started researching the history of medical ethics, I recognized the struggle against utilitarianism with the force of lived experience. Whether my identification with the protagonists on both sides of this story strengthens or distorts the work is up to the reader to decide.

This is not the place to go into all of R. M. Hare's arguments, but his central claim so perfectly exemplifies the sweet promise and bitter reality of utilitarianism that I am compelled to put it on the table. Hare proposes that moral thinking is enacted at two different "levels." At an everyday level, we operate according to a set of absolute rules: do not kill, lie, cheat, steal, break promises, or hurt people. These, he suggests, constitute our "moral intuitions," and they work well—in that they promote human happiness—under most circumstances. But, he cautions, we all run into circumstances when two of the rules conflict. Then what should we do? His answer is that we must drop moral rules in favor of utilitarian calculation. When a promise to meet a cousin for tea conflicts with a call to help a friend in trouble, we must weigh the harms and benefits and choose the alternative that prevents the most suffering. We should incur the small social discomfort caused by canceling the tea in order to prevent the pain engendered by abandoning a friend in need. Instead of relying on unvarying rules of conduct, the good utilitarian figures out what effect his actions will have on the balance of pain and pleasure in the world.

Hare acknowledges that these dispassionate calculations can come across as emotionally repellent. When we have to break the rules of right conduct in order to promote the greater good, it can trouble our deepest intuitions about right and wrong. But, he argues, most criticisms along these lines take the form of thought experiments with no relation to the real world. Here he is rehearsing one of the hackneyed scenarios designed to incite our repugnance at the specter of greater-good reasoning:

> There are in a hospital two patients, one needing for survival a new heart and the other new kidneys; a down-and-out who is known to nobody and who happens to have the same tissue-type as both the patients strays in out of the cold. Ought they not to kill him, give his heart and kidneys to the patients, and thus save two lives at the expense of one? The utilitarian is supposed to say that they ought; the audience is supposed to say that they ought not, because it would be murder.[24]

Hare's answer is simply to point out that the case is implausible: the doctors cannot know that the homeless man has no one to mourn him, nor could they recruit all the other players in this sordid drama to keep their actions a secret. There is no *realistic* way to make murdering the indigent patient the right utilitarian thing to do.[25]

Hare characterizes his two levels of moral thinking as "critical" and "intuitive"—critical calculations of consequences on the one hand, versus intuitive adherence to rules on the other. The intuitive (rule-keeping) level he dubs the "prole" (after George Orwell in *1984*); the critical (utilitarian) level he calls the "archangel." He then uses this terminology to raise an important issue: "When ought we to think like archangels and when like proles?" Admitting that "there is no philosophical answer to the question," he cheerfully throws the reader back on her own self-knowledge: "It depends on what powers of thought and character each one of us, for the time being, thinks he possesses. We have to know ourselves in order to tell how much we can trust ourselves to play the archangel without ending up in the wrong Miltonic camp as *fallen* arch-angels."[26]

In this respect, history provides resources that philosophy cannot. "Fallen archangels" throng the pages of this book, not as cardboard cut-outs inhabiting implausible medical thought experiments, but as real sci-entists caught up in the great social experiment of utilitarian medicine, as it was actually enacted. There are "proles" in this history, too—albeit not in quite the way Hare meant by his use of the term. The fictitious "down-and-out" in his thought experiment stands in for the tens of thousands of poor and institutionalized people who were experimented on in the name of utility. The next chapter culminates in the hearings at which a handful of these ordinary people bore witness to the archangelic crimes of science. I hope the story will show how utilitarianism can send scien-tists down the proverbial well-paved road to hell.

1

Trial of the Archangels

NUREMBERG

The story of the archangels' fall begins, appropriately enough, in an infernal time and place. Mandatory informed consent was born amid the ruins of Nuremberg, the ancient Bavarian city where Hitler had held his annual Nazi Party rallies, images of which can still startle and disturb with their atmosphere of overwhelming unity of hateful purpose.[1] Nuremberg had been bombed to rubble by the Allies during the last months of World War II. A fierce ground battle in April 1945 left even more damage in its wake. After the German surrender on May 8, the devastated city found itself in the center of the American zone of occupation.

The question had already arisen of where in the shattered world the Nazi war criminals should be tried. In Nuremberg, the old law court building, a grand and gloomy structure from 1909, was left standing at the end of the war, miraculously intact amid still-smoking wreckage. The symbolism proved irresistible. After a brief wrangle with the Soviets, who wanted to try the Nazis in their territory in Berlin, it was determined that the International Military Tribunal—a cooperative effort involving judges and lawyers from all four occupying powers—would be held in the Nuremberg courthouse. The Palace of Justice, as the building was known,

27

Figure 2. The Palace of Justice in Nuremberg, photographed on October 1, 1946, the day the verdict in the International Military Tribunal was announced, two months before the start of the Medical Trial on December 9. This image of a German courtroom under American occupation represents all the jurisdictional complexities and historical ironies of the undertaking. In order to convict the Nazi doctors of a crime, the American prosecution lawyers had to reconstruct an ethical framework that could plausibly represent prewar German standards of medical practice and to assert that this framework was the ethic of medical science across the whole "civilized world." Kantian moral philosophy was thus transformed into a universal standard against which all medical research would eventually be judged. Courtesy of dpa picture alliance/Alamy Stock Photo.

was hastily refurbished. A wall was removed in the main courtroom to accommodate the anticipated crowds of spectators and reporters. Between November 1945 and October 1946, twenty-one Nazi defendants stood trial for crimes against peace, war crimes, and crimes against humanity, the world's first foray into international law.[2]

As teams of Allied investigators researched the details of the Nazi killing machine, they started hearing horrifying accounts of medical experiments carried out in the camps. In May 1946 a commission was set up to investigate "War Crimes of a Medical Nature." Clearly, these were criminal activities of a very special kind. Meetings of the commission revealed deep divisions among the members on the nature of the wrong that had been done. The question of the scientific utility of the experiments kept muddying the waters. In one revealing exchange, a French representative asked if death had to result in order for scientific research to count as criminal. A spluttering reply to this question from the commission chairman—"We can not discuss what is crime and what is not"—exposed the bafflement of the investigators faced with heinous acts committed in the name of the healing arts.[3] Another four-power trial began to seem unrealistic, and the Americans started to pursue their own case against Nazi medical science. In October 1946 President Truman announced that a second series of war crimes trials would be held in Nuremberg under the exclusive jurisdiction of the United States. The first of the series was to be the Medical Trial.

The American investigators quickly ran into a legal and ethical roadblock. Most of the Nazi experiments had been conducted on prisoners condemned to death. From a strictly legal standpoint, these researches in the death camps might legitimately be compared with studies conducted in America's prisons. In June 1945, for example, *Life* magazine had run a piece celebrating a series of malaria experiments in American penitentiaries, in which eight hundred men had been deliberately infected with the disease using trapped mosquitoes. The article featured the magazine's signature black-and-white photographs, including one of a prisoner lying in bed, eyes closed, face drawn and pale, in whom, according to the caption, the disease had been "allowed to progress considerably."[4] The American lawyers started to worry about the numerous examples of Allied human experimentation that could be presented at the trial as part of the defense's argument. It was a genuinely knotty question. How could the prosecution make a case against the Nazi doctors without catching American research practices in the toils of international justice?

The attorneys cabled the secretary of war, who turned to the surgeon general, who suggested that the American Medical Association was the appropriate body to untangle the mess. The AMA nominated an eminent physiologist by the name of Andrew Ivy, then in his early fifties. As the scientific director of the Naval Medical Research Institute during the war, Ivy had supervised high-altitude studies and seawater studies using conscientious objectors, work that was similar—at least in substance—to some of the investigations undertaken by the Nazis. An energetic defender of animal experimentation, he had also shown that he could be trusted to protect the interests of the American medical profession in the face of fierce ethical criticism. Media-savvy and a proven believer in the value of aggressive experimentation, Ivy seemed the perfect man for the job.[5]

In July 1946 Ivy set off for Washington, DC, where he spent a few days reading up on what was known about Nazi medical crimes. He then flew to Germany and met with a group of German aviation scientists. At the end of the month he traveled to Paris to attend a meeting of the International Scientific Commission on War Crimes of a Medical Nature. Here he swung into action on behalf of the medical research establishment, stressing that "the publicity of the medical trial should be prepared so that it will not stir public opinions against the ethical use of human subjects for experiments." What was needed, he suggested, was a "pragmatic instrument" to be "adopted and publicized for the purpose of contrasting ethical and unethical experimentation." In a report he sketched the outline of such an instrument, in the form of a set of "Principles and Rules of Experimentation on Human Subjects." The first of these principles was "Consent of the subject."[6]

On December 9, 1946, the Medical Trial opened. Twenty-three defendants were accused of barely imaginable atrocities. Their victims had been put in low-pressure chambers to test the effects of high altitude, until they suffocated from lack of oxygen. They were made to stand naked in subzero temperatures, screaming with pain as their bodies froze. They were deliberately infected with malaria and epidemic jaundice. Sections of bone, muscle, and nerve were removed from their arms and legs. Wounds were inflicted on them, exacerbated with ground glass and wood shavings, and infected with gangrene, bacteria, and mustard gas. They were starved and given nothing but chemically treated seawater to drink. They were burned with phosphorus from incendiary bombs. In the most macabre example of all, 112 Jews imprisoned at Auschwitz were murdered and "defleshed" in order to provide material for a skeleton collection, commissioned for the express purpose of displaying the anatomical basis of their racial inferiority.

The proceedings began with a long and stirring speech by the chief prosecutor, a charismatic, handsome, ambitious lawyer by the name of Telford Taylor. Taylor pointed out that "to kill, maim, and to torture is criminal under all modern systems of law." Given the horrendous nature of the crimes, he explained, "no refined questions" of morality needed to detain the tribunal: "Were it necessary, one could make a long list of the respects in which the experiments which these defendants performed departed from every known standard of medical ethics," he observed. "But the gulf between these atrocities and serious research in the healing art is so patent that such a tabulation would be cynical."[7]

Back in the United States, however, the American Medical Association was embarking on exactly the cynical tabulation that Taylor had pronounced unnecessary. Far from being dispensable, these ethical questions were in urgent need of resolution, forced onto the table by the prospect of *all* human experimentation, Allied and Nazi alike, being put in the dock in Germany. On December 10, one day after the opening statements in Nuremberg, the judicial council of the AMA formally adopted a skeleton version of Ivy's guidelines for human subjects research. Buried deep in the small print of the *Journal of the American Medical Association* a couple of weeks later, the following announcement could have been spotted by a particularly assiduous reader:

> In order to conform to the ethics of the American Medical Association, three requirements must be satisfied: (1) the voluntary consent of the person on whom the experiment is to be performed; (2) the danger of each experiment must be previously investigated by animal experimentation, and (3) the experiment must be performed under proper medical protection and management.[8]

In this aborted little paragraph, the consent provision did not even merit a full sentence.

Back in Nuremberg a few weeks later, in January 1947, the trial took the turn the prosecution had feared. The Americans had called as a witness a German psychiatrist who had been persecuted by the Nazis during the war. They hoped that he would attest to the state of German medical ethics before the Third Reich, and thereby establish a baseline of homegrown legal standards against which the defendants' actions could legitimately be judged. The witness duly outlined what he saw as the core of Hippocratic medicine: a reverence for life, which had been grossly dishonored by his compatriots. The leading defense lawyer then proceeded to cross-examine him. After getting the witness to agree that experimentation on prisoners was unethical, the lawyer whipped out the *Life* magazine article about malaria research in American penitentiaries. With dramatic emphasis he read it aloud, describing in detail each of the featured photographs. At the end of this performance he asked the man in the stand if the research sounded ethical. The witness admitted that it did not.

Ivy was in the courtroom during this exchange. He promptly flew back to the United States, where he met with Dwight Green, the governor of Illinois, who had authorized the largest of the prison studies, and suggested that they convene a committee to investigate the experiments. The governor agreed, and in March 1947 he contacted some Illinois citizens to see if they would be willing to serve. In June Ivy went back to Nuremberg, where he testified that the committee had concluded that the prisoners had volunteered for the malaria experiments as freely as any free men. It was a lie. The "Green Committee" had not, in fact, ever convened. Ivy had written the exculpatory report himself.[9]

Following up, one of the American prosecution lawyers asked Ivy if the new AMA rules accurately represented the principles upon which "all physicians and scientists guide themselves before they resort to medical experimentation on human beings in the United States?" Ivy answered, "Yes." "To your knowledge," the lawyer prompted, "have any experiments been conducted in the United States wherein these requirements which you set forth were not met?" Ivy answered, "Not to my knowledge." At this point one of the defense lawyers jumped in, objecting that the American Medical Association ethical guidelines had been adopted in 1946 and had "no retroactive force." He was overruled. It was allowed to stand that voluntary consent of research subjects was always—and had always been—sought by American and Allied researchers, in conformity with the ethics of the whole "civilized world."[10]

In vain did the defense suggest that the tribunal "study the many experiments performed all over the world on healthy and sick persons, on prisoners and free people, on criminals and on the poor, even on children and mentally ill persons," to establish the fact that their Nazi clients had been adhering to internationally approved ethical standards.[11] In vain did they submit a document detailing research conducted in Allied countries, involving a total of approximately eleven thousand human subjects. As well as the eight hundred Illinois prisoners infected with malaria, the memo listed the controlled malnutrition diets inflicted on prisoners in New York State that gave seven of them pellagra, the cholera trials in India in which a control group was left untreated resulting in scores of deaths, the administration of poisons to condemned prisoners, the deliberate infection of conscientious objectors with hepatitis, the six cancer patients who died after experimental cooling therapy, and many more.[12] The researchers in charge of these experiments, the defense pointed out, "gained general recognition and fame; they were awarded the highest honors; they gained historical importance."[13] Why, then, were their Nazi clients being treated as criminals for the same kinds of activities?

One of the lawyers for the accused challenged Ivy to ponder a thought experiment in which a plague-ridden city could be saved with the sacrifice of a single condemned prisoner. Asked whether he was "of the opinion that the life of the one prisoner must be preserved even if the whole city perishes," Ivy answered, "Yes."[14] He could see where this line of questioning was leading—if it was permissible to use one prisoner to save a city, then it was permissible to experiment on prisoners in German concentration camps to save an army—and he was not going to put so much as a tentative toe on the slippery slope of utilitarian arguments. In the context of the trial, the perilous elasticity of the concept of the greater good called for a principle that enshrined the absolute rights of every individual over the calculable benefits to the collective.

Accordingly, in August 1947, when the judgment in the Medical Trial was handed down, it sanctified the principle of informed consent, strengthening and formalizing it as the first and preeminent provision of the Ten Commandments of experimentation ethics now known as the Nuremberg Code:

> The voluntary consent of the human subject is absolutely essential. This means that the person involved should have the legal capacity to give consent; should be so situated as to be able to exercise free power of choice, without the intervention of any element of force, fraud, deceit, duress, over-reaching, or other ulterior form of constraint or coercion; and should have sufficient knowledge and comprehension of the elements of the subject matter involved as to enable him to make an understanding and enlightened decision. The latter element requires that before the

acceptance of an affirmative decision by the experimental subject there should be made known to him the nature, duration, and purpose of the experiment; the method and means by which it is to be conducted; all inconveniences and hazards reasonably to be expected; and the effects upon his health or person which may possibly come from his participation in the experiment.[15]

The first commandment of the Nuremberg Code is replete with historical irony. In 1931, on the eve of Hitler's rise to power, a set of guidelines issued by the Health Council of the German Weimar Republic included the injunction: "A new therapy may only be administered if the person concerned or their legal proxy gives unambiguous consent on the basis of an appropriate explanation."[16] At the Nuremberg Trial, Ivy had appealed to these 1931 guidelines as evidence for the universality of medical ethics.[17] But mandatory informed consent was far from universal. It was, in fact, quintessentially German, springing from a concept of transcendent human dignity found in the moral philosophy of Immanuel Kant.[18] No other nation had seen fit to formalize the principle of consent before the war.

There is little doubt that the framers of the Nuremberg Code, including Andrew Ivy, saw consent as the proper foundation for medical ethics. For most Allied researchers, however, there was another ethical system that trumped it. The principle of societal benefit was so deeply rooted in the practices, goals, and achievements of scientific medicine that mandatory informed consent was regarded as not just unnecessary but as actively burdensome and counterproductive. The rights of the patient or research subject did not figure in this ethos. It was the scientist who enjoyed rights, including the right to distribute the burdens and benefits of the research enterprise as he saw fit.

A representative example of Allied research ethics was brandished by the defense lawyers at the Nuremberg Medical Trial as one of their prize exhibits. It was a 1946 memo from British physiologist Robert McCance to the president of the Northern Rhine province of Germany, asking to be informed if any children were born in hospitals in the region with "abnormalities, which will make unlikely or impossible that the children will survive longer than a short time," as he wished "to make some experiments on those children which will give them no sorts of pains."[19]

McCance was a distinguished scientist who had risked his life in

World War I flying observation aircraft from perilously short launching platforms on warships. A physiologist, pediatrician, and nutritionist, he held the chair of experimental medicine at Cambridge University. After Nuremberg, he was moved to lay out his views on medical ethics, presumably as a result of having been cited at the trial. The article, published in the *Proceedings of the Royal Society of Medicine* in 1951, constitutes a concise summary of the prevailing ethical standards in Allied countries. Cheerfully admitting that investigators and physicians have different goals and priorities, McCance described how he enrolled people in experiments. He explained that he had "never had much difficulty in obtaining the cooperation of normal subjects" and remarked, "It is always a great help to have made the same experiment on oneself first and to be prepared if necessary to do so again." The ethos of self-experimentation was a guarantee of the researcher's integrity as well as of the safety of the experimental drug or procedure.

So far so good, but this was only the start of McCance's manifesto. While healthy people, such as laboratory technicians (his ideal subjects), should only be experimented on with their cooperation and consent, patients in hospital were another matter: "Many experiments, even quite elaborate ones, can be made on patients, within the therapeutic routine of the hospital . . . without anyone thinking anything of it," he declared. Asking permission of patients for these investigations was quite unnecessary, an axiom that McCance stressed by repeatedly putting the word "permission" in quotation marks.

The whole scheme rested on an assumption of trust in the medical profession: "the patient trusts the staff of the hospital, and the investigator, knowing this, usually dispenses with the formality of asking for 'permission.'" At the end of the article McCance declared that medical researchers "have a right to expect the fullest cooperation from their medical colleagues, from nurses and other assistants, from hospital managements, from patients and relatives and from the community at large."[20] At the time, the National Health Service was just two years old, and this philosophy of the greater good was a perfect fit with the communitarian ideals of postwar British socialism.

McCance's views were widely shared by researchers in all the Allied nations.[21] In most democratic countries experimenters balanced harm to research subjects against benefit to society, using a variety of approaches to enroll people, including, in a subset of cases, soliciting their consent. For the purpose of recruiting colleagues, a gentlemanly agreement between equals sufficed. In this context, self-experimentation was an important aspect of building trust. In larger trials, healthy subjects were often drawn from prisons, and here inducements such as small sums of money, better food in the infirmary, a break from routine, free cigarettes, or favorable letters to the parole board were enough to secure eager recruits. In the case of research on hospital patients, experimentation was discreetly folded into the routines of testing and therapy without the subjects' knowledge. As for people in institutions for the mentally ill and the disabled (never mentioned by McCance, despite his 1946 appeal for experimental access to disabled infants), the terrible conditions of life in such places were seen as justifying the negligible added burden of serving as unwitting guinea pigs. Consent was far from an absolute moral imperative; it was just one among a number of procedures and strategies for getting subjects to participate in medical research. In all cases, the scientist acted as the trusted guardian of the highest societal benefit.

Unfortunately for Andrew Ivy, this unsystematic mixture of intuition, etiquette, and tradition defied any sort of crisp definition in the context of the legal battle in Nuremberg. The lawyers for the Nazis were able to identify the prevailing ethos with stark utilitarianism, and then line it up with all nontherapeutic experimentation done in the name of medical science, including the activities of their clients. This forced the prosecution to enshrine unfettered *voluntariness* as the prime directive of research ethics, a standard that did not really apply to anyone in the research enterprise, except, perhaps, the self-experimenting scientist and the handful of coworkers he might persuade to join him at the very earliest stages of testing a new drug or procedure.

ARCHANGELIC RESEARCH AFTER NUREMBERG

After the Nuremberg Medical Trial, human experimentation in America developed in two opposing directions. On the one hand, some agencies and institutions began to formalize consent procedures along the lines of the Nuremberg Code. The general manager of the Atomic Energy Commission, for example, issued a document outlining the informed consent requirement in 1947, immediately after the conclusion of the trial. Implementation, however, was another matter, and most experimenters within the military seemed unaware of the existence of this requirement.[22] The National Institutes of Health adopted a policy based on the Nuremberg Code for their new Clinical Center.[23] Successive editions of a textbook called *Doctor and Patient and the Law* paid increasing attention to consent issues, with the 1956 edition containing a discussion of the proceedings at Nuremberg. In 1959 an organization cofounded by Ivy sponsored a conference called "The Legal Environment of Medicine," at which the Nuremberg Code was discussed.[24]

On the other hand, most researchers seem to have either ignored the Nuremberg Code or regarded it as irrelevant to the beneficent practices of American medicine. In fact, the late 1940s and 1950s were characterized by the scaling up and intensification of research conducted without the consent of the subjects.[25] Especially in the United States, the research ethics outlined by Robert McCance gave way to an even more starkly utilitarian approach. Larger numbers of trial participants were enrolled in riskier studies. National security was invoked to justify secrecy. Self-experimentation by scientists often dropped out of the picture, replaced by more calculated selection of "unfortunates" to serve as research material.

The shift toward a starker utilitarianism can only be understood against the backdrop of World War II. Extreme hazards faced by soldiers in combat had produced an ethos of sacrifice for the common good. Human medical experimentation during the conflict was understood as an "admirable element in the home front effort."[26] The metaphor of battlefield sacrifice continued into the early years of the Cold War. A famous 1945 polemic, *Science, the Endless Frontier*, called for a "War against Disease."[27] In the 1950s newspaper and magazine stories extolled the heroism of the prisoners, soldiers, conscientious objectors, and medical students who served as "human guinea pigs" during the Korean War.[28] Moreover, the mass production of penicillin, initiated as part of the war effort, created a sense of limitless therapeutic optimism. Penicillin was so stunningly effective that the public was willing to support all kinds of research in the hopes of more such miracle cures. These new therapeutic powers and potentials shifted medical research in the 1940s and 1950s in an even more utilitarian direction, the Nuremberg Code notwithstanding.

One study that has recently come to light can stand as an example of utilitarian research ethics in the age of penicillin.[29] A public health researcher by the name of John Cutler was working in the US Public Health Service Venereal Disease Research Laboratory on Staten Island during the war when penicillin first began to be mass-produced. He had observed firsthand the resulting breakthrough in the treatment of syphilis and gonorrhea: the new drug could effect a cure within one week. After taking penicillin, however, many people quickly got reinfected. Cutler noted that some older studies of syphilis had seemed to indicate that a slower course of treatment with the traditional remedies of arsenic derivatives and mercury gave time for the body to develop lasting immunity to the disease.[30] The phenomenon of rapid cure followed by reinfection seemed to him to merit "restudy of experimental syphilis in the light of . . . penicillin therapy."[31] Ultimately, he wanted to see if penicillin, alone or in combination with the older therapies, might be used as a prophylactic against sexually transmitted diseases, in keeping with the priorities of a War Department much preoccupied with the problem of venereal disease among the troops.

It happened that Juan Funes of the Guatemala Public Health Service had just arrived at the laboratory as a research fellow. According to Cutler, it was Funes who suggested "carrying out carefully controlled studies [of syphilis] in his country." Funes explained to Cutler that prostitution had been legalized in Guatemala, to the extent that sex workers were even allowed to ply their trade in penal institutions. Between them, the two men came up with a plan in which a prostitute who tested positive for syphilis or gonorrhea would be paid by the public health department to go to a prison and offer her services to "any inmate who desired to utilize her at no cost to himself."[32] Having created an experimental population of infected prisoners, they would then test penicillin and other therapies for their prophylactic potential.

The proposal was well received, and Cutler was awarded a generous sum to run the investigation. Things did not, however, go smoothly. When his team embarked on a preliminary diagnostic program in a prison in Guatemala City, they immediately ran into a series of roadblocks. The startlingly high percentage of positive results cast doubt on the reliability of the tests, and most of the prisoners were resistant to having their blood taken. The study moved to working with soldiers in the Guatema-

lan army and patients in a psychiatric hospital. It turned out that rates
of infection from sexual contact with prostitutes were much lower than
expected. The researchers abandoned their original study design and
moved to what they called a "direct inoculation" technique. This involved
forcibly introducing pus from acute cases of syphilis and gonorrhea into
the mucous membranes of research subjects. By the end of two years,
1,308 people—prisoners, sex workers, soldiers, orphans, and psychiatric
patients—had been intentionally infected. Of these only 678 seem to have
received treatment, some with the old, pre-antibiotic therapies. Cutler's
records note 83 deaths during the course of the deliberate exposure experiments, most of which he attributed to tuberculosis.[33]

One thing that comes across clearly in the paper trail from the study
is Cutler's opinion of the hopelessness of his experimental subjects' lives.
A long passage in his report describes the asylum inmates' "pathetic
anxiety to participate," which he attributed partly to the cigarettes that
were given out with every procedure and partly to their being "starved
for attention and recognition as individuals."[34] This paternalistic attitude
to the desperate conditions of the time and place was also used to secure
the Guatemalan government's cooperation. Control over supplies of penicillin, for example, became a card that Cutler played "to build goodwill."[35]

Most of the researchers involved in this study were aware that it
would not withstand public scrutiny. The use of psychiatric patients, for
example, stirred up anxiety in one of the supervisors of the experiment,
who commented that "insane people . . . cannot give consent" and so if
"some goody organization got wind of the work, they would raise a lot of
smoke."[36] Cutler himself suggested that it be kept secret. It was, nonetheless, widely condoned and supported at the highest levels of the research
establishment. The US surgeon general, Thomas Parran Jr., is reported to
have followed it with fond attentiveness. One army researcher wrote that
Parran "is very much interested in the project and a merry twinkle came
into his eye when he said, 'You know, we couldn't do such an experiment
in this country.'"[37]

Working in an age of variable and fluid ethical standards, driven by
the ethos of the greater good, and armed with unprecedentedly effective
new therapies, American researchers found it easy to justify the deliberate infection of human subjects with a deadly, communicable disease. A
flowery letter from the medical director of the Guatemalan prison where

Cutler began the study suggests the rapturous reception that his work enjoyed:

> It is a privilege for us to manifest you, by means of these lines, our everlasting gratitude, which will remain for ever in our hearts, because of the noble and gentlemanly way with which you have alleviated the sufferings of the guards and prisoners of this penitentiary. You have really been a philanthropist, your disinterestedness, your constancy, are evident samples of your nobleness . . . etc.[38]

Who wouldn't think of themselves as a bit of an archangel after getting a letter like that?[39]

At the other end of the secrecy scale from the Guatemalan experiment lies the 1954 American trial of the polio vaccine, undertaken in a blaze of publicity amid the polio epidemics of the early 1950s. When Jonas Salk developed his killed-virus vaccine, he was convinced enough of its safety that he first injected it into himself and his children, after boiling the syringes on the stovetop in their kitchen. He had earlier suggested that they look into "institutions for hydrocephalics and other similar unfortunates" to find test subjects, and in the end he ran the first human trials for safety and efficacy at a "home for crippled children" (with many polio cases) and a "school for the retarded and feeble-minded."[40] From start to finish, Salk obeyed the highest prevailing norms of the profession—extensive animal testing, self-experimentation, and experimentation on his own children, followed by very careful, small-scale trials on children in institutions, some of whom were polio sufferers.

After the tests in the children's homes showed promising results, a full trial was undertaken. Public trust had been established by Salk's willingness to experiment on himself and his children, and epidemic conditions had shifted the calculus decisively in the researchers' favor. On the form that parents signed to enroll their children, the phrase "I give my permission" was changed to "I hereby request," signaling that the trial was a privilege to which they were petitioning for access, as opposed to a hazard that they were willing to undertake. (This same shift would occur in the very different context of the AIDS epidemic, as explored in chapter 6.) By the time the clinical trials were complete, more than 1.3 million children had received either the vaccine or a dummy injection, a testament to the American public's support for this kind of medical research. No one blinked at the vaccine having been tested on institutionalized children. Millions of parents signed the forms requesting access to the trial. When its successful conclusion was announced on April 12, 1955, the much-vaunted "War against Disease" was seen to have won a great victory. One author, then a fourth grader, characterized it as "another V-J Day—the end of a war," and recalled the sound of "car horns honking and church bells chiming in celebration. We had conquered Polio."[41]

THE END OF THERAPEUTIC OPTIMISM

Salk's polio vaccine trial marked a high point of public trust in the medical research enterprise, but in the early 1960s the social contract that held utilitarian medical research together began to fracture. In November 1961 it emerged that thalidomide, a sedative developed by the German pharmaceutical company Chemie Grünenthal, caused severe birth defects. The drug was licensed in forty-six countries under dozens of different brand names. When patients began to complain of severe and sometimes irreversible nerve damage, Grünenthal did what it could to cover up the reports, redoubling efforts to promote the product. After a circular went out from the company assuring physicians that it was harmless for pregnant women, doctors began to prescribe thalidomide for morning sickness. As a result, between eight and ten thousand children were born with extremely disabling and stigmatizing birth defects, including the characteristic lack of arms or legs, with feet emerging from the hips and hands from the shoulders. Just one or two tablets taken at the wrong time during pregnancy were sufficient to cause complete malformation of whatever bodily system happened to be developing at the time. Babies were born without ears, without anuses. Internal organs were also affected. Many of the children died. Infants were abandoned; parents had breakdowns; thousands of lives were shattered.

One of the directors of Chemie Grünenthal was Heinrich Mückter, a chemist who had been a high-ranking Nazi medical officer in Poland.[42] In pursuit of corporate profit, Mückter engaged in a systematic cover-up of the dangers of the drug. When the complaints of nerve damage began to surface, he did all he could to suppress them. Even when confronted with evidence about the birth defects, he refused to pull the product from the shelves. When the case came to court in 1968, he declared himself outraged to be held responsible for the safety of the product. On the other side of the courtroom stood the pediatrician and medical researcher who first made the connection between birth defects and the drug.[43] The judges ruled that this prosecution witness was insufficiently impartial. Because his sympathies so clearly lay with the disabled children, his testimony was struck from the record. As a result of this legal gagging, the company was able to settle out of court for derisory sums to the thousands of German victims.

In the United States, the drug never received a license. A company called Merrell had bought the US rights, but its application for a license was turned down by a Food and Drug Administration scientist called Frances Kelsey. Kelsey had been a professor of medicine at the University of South Dakota before taking a job at the FDA. Her suspicions were aroused because the drug's pronounced sedative effect in humans was not replicated in animals, leaving open the question of other unanticipated side effects in humans. Shortly afterward, the first reports of nerve damage caused by the drug came across her desk. Kelsey knew from her wartime research on rabbits that compounds that irritated the adult mammalian nervous system could cause fetal deformities. At the time, the FDA was required to hand down a decision on a new drug within sixty days; otherwise its approval was automatic. Resisting pressure from company executives, Kelsey applied for extension after extension, until finally the news broke about the birth defects and Merrell's application was hastily withdrawn.

Too late. It emerged that American doctors had given out tens of thousands of samples of the drug to patients, without informing them that it was an unlicensed experimental product or following up with them about the effects of the treatment. Efforts to trace the two and half million tablets that Merrell had distributed in the United States were thwarted by inadequate recordkeeping. In the end, seventeen American children were born with birth defects. Hearings were held. Stories of other drugs that produced severe side effects—these ones American-made—were read aloud in Congress. American consumers stopped buying pharmaceuticals, and the price of shares in the sector began to plummet, losing anywhere from 10 to 60 percent of their value. Even sales of aspirin dropped.[44]

A couple of years before the news broke about thalidomide's deadly effects, Senator Estes Kefauver, a Democrat from Tennessee, had joined battle against price fixing in the pharmaceutical industry. The thalidomide disaster boosted Kefauver's campaign, endowing it with urgency and enlarging its scope. Despite the clear and present danger of regulatory laxity, however, Kefauver himself balked at the suggestion that doctors be forced to get their patients' permission before giving them unlicensed drugs because that would be "an infringement of the physician's rights."[45] It was two Democrats irritated by obstructionism on the part of their Republican colleagues who decided in a committee meeting to strengthen the regulations on drug testing, "*going so far* as to require the consent of the patient."[46]

The Kefauver-Harris Drug Amendments to the Federal Food, Drug and Cosmetics Act were signed into law by President Kennedy in October 1962. These amendments contained, for the first time, the informed consent mandate in a piece of federal legislation. Researchers using "drugs for investigational purposes" were required to inform their human subjects that the substance was experimental and to obtain their consent. This is sometimes cited as a crucial turning point in the development of contemporary medical ethics, but the 1962 law contained a legal loophole big enough to pass just about anything through. An exemption was provided for cases in which researchers "deem it not feasible or, in their professional judgment, contrary to the best interests of such human beings," effectively leaving the whole business still at the discretion of the scientists rather than the subjects.[47]

In 1962 an NIH official was asked to investigate the "moral and ethical aspects of clinical investigation." The resulting report concluded, "Whatever the NIH might do by way of designing a code or stipulating standards for acceptable clinical research would be likely to inhibit, delay, or distort the carrying out of clinical research."[48] Just as Robert McCance had asserted in his 1951 article on medical ethics, it was physicians and researchers who enjoyed inalienable rights in the context of scientific medicine, appropriate to their status as philanthropists laboring for the greater good. No inflexible moral code was worth the price of inhibiting, delaying, or distorting this supremely valuable work.

COMPARISONS WITH NAZI MEDICINE

Even in the face of the thalidomide tragedy, the American government did not see the need for wholesale reform, contenting itself with tightening up a system that in most people's view basically worked. The following year, however, a scandal broke on the domestic front, prompting, for the first time since the Nuremberg Medical Trial, a direct legal comparison between American utilitarian medicine and fascist human experimentation. A researcher by the name of Chester Southam had been pursuing a theory about the immunology of cancer. He hypothesized that the disease was caused by a defect in immune response. Accordingly, he wanted to see if people with cancer took longer to reject foreign cells than people without the disease. He had a relatively new tool at his disposal, the first immortal human cell line, derived in 1951 from a dying woman's cervical cancer tissue.[49] His plan was to inject these cells into the patients' arms or legs and see how long it took for a "subcutaneous nodule" to form and then to disappear. To test his theory, he calculated that he needed three different groups of subjects: cancer patients, healthy subjects, and sufferers from noncancerous terminal illnesses. Comparing the rates of rejection between these three groups would tell him if cancer patients had defective immune systems.

The first group—cancer patients—he found at his home institution, the Sloan-Kettering Institute for Cancer Research. Southam told these patients that he was injecting them with "cells," not mentioning that the cells in question were cancerous. The second group—the healthy volunteers—he recruited at the Ohio State Penitentiary. These men were given the benefit of full informed consent and hailed in the local press as heroes. Embarking on the third phase of the research in 1963, Southam approached administrators at the Jewish Chronic Diseases Hospital in Brooklyn to request access to a group of terminal noncancer patients. The director of medicine agreed. He instructed his staff to give the injections, specifying that they should not tell the patients that the "skin test" consisted of cancer cells. Three doctors refused. When the experiment went ahead anyway, the doctors resigned.

Southam was only proposing to do exactly what everyone else did—to integrate experimentation into the therapeutic routines of a hospital without asking patients' permission. He was following the hitherto uncontroversial moral code laid out in 1951 by Robert McCance. But times had changed. The trial of Adolf Eichmann in Jerusalem had aired live on television for sixteen weeks in 1961, raising awareness of Nazi atrocities to unprecedented heights. It was an inauspicious moment to ask Jewish physicians to violate the Nuremberg Code. At a meeting of the hospital's board of directors, amid heated argument about what exactly patients had been told about the injections, the chairman of the board abruptly adjourned the proceedings, invoking "the Nuremberg Trials in which Nazi doctors were found guilty and some hanged . . . for using human beings for experimental purposes without their informed consent."[50]

After the board meeting one of the members, a lawyer, sued the hospital for access to the study records. In his petition, he stated that Southam was injecting cancer cells into noncancerous patients "for the purpose of determining whether cancer can be induced," a slight but devastating misrepresentation of the research objectives, and a harbinger of the inflamed rhetoric that would develop around the case.[51] While the injections did produce "nodules," Southam was confident that the cells would be rejected; he just wanted to see if the noncancer terminal patients rejected them quickly, like his healthy subjects from the penitentiary, or slowly, like his cancer patients at Sloan-Kettering. Another affidavit in the lawsuit referred to "acts that belong more properly in Dachau."[52]

Southam did nothing to dispel the aura of hubris and arrogance around the case. He was interviewed by a reporter for *Science* magazine, who asked him why he did not first try out the injections on himself. He replied, "Let's face it, there are relatively few skilled cancer researchers, and it seemed stupid to take even the little risk."[53] Later, at his licensing hearings, he was asked about this statement. He replied that he did not remember saying those exact words, but agreed repeatedly that "the philosophy is correct." Southam's refusal to self-experiment revealed a subtle shift in the moral norms of the research enterprise in the age of penicillin. Now that medicine had proved how miraculously effective it could be, self-experimentation violated the strict utilitarian calculation that weighed risk to subjects against their potential to contribute to society. A prisoner or terminally ill hospital patient was one thing. A skilled cancer researcher was something else, a commodity altogether too valuable to be risked on the altar of "false heroism," as Southam put it.[54]

But if medical utilitarianism had hardened, so had legal opposition to such flagrant moral elitism. In 1965 the attorney general of New York State conducted a hearing to determine whether to suspend Southam's license. He characterized the study as "in no way therapeutic . . . an experiment relating to cancer research which had as its ultimate intention the benefit of humanity." In the context of the hearings, the "ultimate intention to benefit humanity" was framed as a ruthless objective that made it "incumbent upon the respondents to have seen to it that ALL information connected with the experiment was given." Against the logic of the greater good and researchers' rights, the attorney general offered the ideal of universal human rights: "Every human being has an inalienable right to determine what shall be done with his own body . . . a right to know what was being planned . . . a right to be fearful and frightened and thus say NO."[55] After a decade of the civil rights movement, doctors could no longer float above the law, and patients could no longer languish outside it.

THE END OF MEDICAL UTILITARIANISM

In 1965 Chester Southam's research was exposed as just one among many such studies conducted on unwitting human guinea pigs. On March 23, the crusading anesthesiologist Henry Beecher summoned members of the press to hear a paper he was giving at a two-day symposium on clinical research. According to the *New York Times*, Beecher's paper, "Ethics and the Explosion of Human Experimentation," cited twenty "breaches of ethical conduct" by American clinical researchers who used "hospitalized patients, military personnel, convicts, medical students, and laboratory personnel for experiments in which the subjects are not asked for their permission."[56] Beecher described Southam's cancer study in a terse two sentences. Other examples included the twenty-three charity patients who died when the treatment for typhoid fever was experimentally withheld from them, the fifty healthy inmates of a children's center who were given repeated doses of a drug that caused abnormal liver function, and the eighteen children undergoing heart surgery who received unnecessary experimental skin grafts. For reasons of professional solidarity and discretion, Beecher did not identify the scientists in question, but he hinted that other such cases would be easy to find.

By the time the paper was published in the *New England Journal of Medicine* the following year, the number of cases Beecher presented had risen to twenty-two.[57] In Beecher's words, the problem was *"experimentation on a patient not for his benefit but for that, at least in theory, of patients in general."*[58] He had nailed it. Experimenting on unwitting patients for the benefit of "patients in general" was unethical. The time was ripe, and the paper made a huge impact. The serene assumption that a researcher could sneak in a few experiments on hospital patients without anyone caring was about to be vaporized in a very public firestorm.

Beecher's sensitivity to this question had a personal dimension. In the early 1950s he had directed an army-sponsored study at Harvard of the truth-serum properties of hallucinogens, in which the subjects were enrolled without their consent.[59] By 1959, however, he had repented of such work and published a report on human experimentation, in which he remarked that "the recent Hitlerian acts" had made responsible investigators "wary of such phrases as 'for the good of society.'"[60] In 1959 his attack on medical utilitarianism was part of a gentlemanly debate within the medical profession; by 1965 his exposé of ethical breaches in medical research detonated like rhetorical dynamite.[61] In its wake dozens of articles and books appeared on the problem of human experimentation, some by physicians and scientists, others by lawyers, theologians, philosophers, and social scientists. The days of the American medical profession's self-legislation were effectively over.

In July 1966, responding to the thalidomide tragedy, Southam's cancer cell experiments, and Beecher's exposé, the National Institutes of Health promulgated guidelines for all federally funded research, mandating that institutions receiving grants create "Review Boards" charged with documenting informed consent. The NIH informed-consent procedure for healthy volunteers, however, fell far short of the standard of voluntariness that had been set at Nuremberg. The explanation for scientists shrugged that "it is not possible to convey all the information to the subject or patient upon which he can make an informed decision."[62] A pamphlet for subjects laconically explained, "You will be asked to sign a statement indicating that you understand the project and agree to participate in it. If you find your assigned project to be intolerable, you may withdraw from it."[63] Setting intolerability as the standard for withdrawal hardly constituted a decisive blow for individual self-determination. Despite the scandals and tragedies, the first American foray into research oversight still fell far short of the Nuremberg Code.

SOCIAL WORTH AND HUMAN DIGNITY

After Beecher's 1966 exposé, there was a veritable explosion of activity, including congressional hearings on the social implications of new medical technology and the founding of dedicated institutes of medical ethics. With this rapid expansion of the field of debate, theologians were among the first to be called upon to weigh in on the issues.[64] The most influential of these was the Princeton professor of religion Paul Ramsey, who had published a series of works examining nuclear deterrence, pacifism, and the concept of a just war. As a theologian attuned to weighty contemporary topics, he was an obvious candidate when the Divinity School of Yale University decided to devote its 1969 lecture series to medical ethics. To prepare for his talks, Ramsey arranged to spend the spring semesters of 1968 and 1969 among the medical faculty of Georgetown University. In 1970 he published his Yale lectures on medical ethics under the title *The Patient as Person*, destined to become his most celebrated work.

The Patient as Person opens with the full text of the first provision of the Nuremberg Code. This is followed by a series of reflections on everything contemporary medical ethics was not: "Medical ethics has not its sole basis in the overall benefits to be produced. It is not a consequence ethics alone . . . not solely a teleological ethics . . . not even an ethics of the 'greatest possible medical benefits of the greatest possible number.'" To the repetitive cadences of this list, Ramsey added another element. Medical ethics, he claimed, was "not solely a benefit-producing ethics *even in regard to the individual patient.*"[65] The anti-utilitarian thrust of informed consent had now reached beyond the research subject to the ordinary patient, who was free to refuse the therapeutic interventions of his or her physician, however benevolently intentioned.

For Ramsey, the trumping of what he called "consequence ethics" by individual choice was replete with spiritual significance. The existence of a moral terrain beyond utilitarianism could only be explained by appealing to the theological sanctity of the individual human spirit.[66] This came across particularly clearly in his discussion of the work of Seattle's

"God Committee," a group of citizens called together in the early 1960s to determine which kidney patients would get access to dialysis treatment, which was then still new, expensive, and scarce. In November 1962 *Life* magazine ran a lengthy article about the agonizing decisions made by this "small, little-known group of quite ordinary people" as they selected who would live and who would die. After excluding as many candidates as possible on practical grounds (no one at home to help them, etc.), the seven committee-members—"a lawyer, a minister, a banker, a housewife, an official of state government, a labor leader and a surgeon"—began, hesitantly, to focus on other criteria. Did the applicant have dependents? Was he the sole breadwinner? Who had "been most provident"? What about "character and moral strength"? Who were "the men with the highest potential of service to society"?[67]

In 1962, when the article was published, this was just a human-interest story. Over the course of the decade, as the anti-utilitarian thrust of medical ethics became ever more vehement, the Seattle committee's appeal to social-worth criteria metamorphosed into a scandal. In 1968 a lawyer and a surgeon cowrote an article in a law journal decrying "how close to the surface lie the prejudices and mindless clichés that pollute the committee's deliberations . . . a disturbing picture of the bourgeoisie sparing the bourgeoisie."[68] As the social hierarchies and certainties of the 1950s were subjected to radical critique, utilitarian judgments of worth seemed increasingly superficial, empty, and corrupt.

In *The Patient as Person*, Ramsey used gentler Christian terms to make the same point, suggesting that "an extension of God's indiscriminate care into human affairs requires random selection and forbids god-like judgments that one man is worth more than another."[69] Other theologians and philosophers followed suit. A philosophical discussion of the God Committee published in 1970 by the Quaker ethicist James Childress argued that the "utilitarian approach . . . dulls and perhaps even eliminates the sense of the person's transcendence, his dignity as a person which cannot be reduced to his past or future contribution to society." Like Ramsey, Childress suggested that the only right way to choose the beneficiaries of scarce medical resources was by means of random selection. "God might be a utilitarian," he admitted, "but we cannot be."[70] In whatever idiom it was clothed—human rights, sanctity, personhood, dignity, transcendence, or autonomy—the unit of moral reasoning that emerged out of the human experimentation scandals of the late 1960s was the free, sovereign, inviolable individual rather than the social collective.

TUSKEGEE HITS THE HEADLINES

Meanwhile, the Tuskegee Study of Untreated Syphilis persisted, hiding in plain sight of anyone who cared to read the published reports. In the mid-1950s, Count Gibson, a community health activist, wrote to one of the study's directors after seeing him lecture about it, questioning the premise of leaving hundreds of people untreated for a dangerous disease: "I am gravely concerned about the ethics of the entire program. . . . There was no implication that the syphilitic subjects were aware that treatment was being deliberately withheld." The researcher replied that he had originally been troubled, but "after seeing these people, knowing them and studying them and the record, I honestly feel that we have done them no real harm and probably have helped them in many ways."[71] In 1964 Irwin Schatz, a cardiologist from Detroit, well aware of the debates about the Nuremberg Code, read one of the published papers on the study and wrote a letter of protest to the Public Health Service. He did not even receive a reply.[72]

In 1965 the study came to the attention of the PHS employee who would blow the whistle on the whole enterprise and finally bring medical utilitarianism to its knees. The earlier critics of the study had hailed from the left of the political spectrum; Peter Buxtun's passion for freedom was of a more conservative stripe. According to historian Susan Reverby, he was a "libertarian Republican, former army medic, gun collector, and NRA member with a bachelor's degree and some graduate work in German history," who had taken a job with the PHS tracking gonorrhea and syphilis patients and their sexual contacts in San Francisco. He had come to America as an infant when his family fled the Nazi occupation of Czechoslovakia in 1939. Living in exile from fascism and then from communism had given him a deep aversion to anything that smacked of totalitarian government. A die-hard maverick, liable to go off on tangential rants in his monthly reports, Buxtun could be an irritant to his supervisors, but he was superb at his job, tracking down sexual partners with the same determination that he would bring to stopping the Tuskegee Study.

In the fall of 1965, Buxtun wandered into the coffee room at work and overheard two older PHS colleagues discussing Tuskegee. One was recounting a case in which one of the unwitting research subjects had got hold of penicillin, whereupon the Centers for Disease Control had jumped down the throat of the prescribing doctor. Buxtun was shocked—how was it possible that he was working day and night to prevent syphilis, and here were his colleagues blithely talking about a PHS study that deliberately left it untreated? He asked for more details and followed up by reading the published material on the study. In his November 1965 report to his supervisor he registered his protest, accusing the Public Health Service of "duplicating the 'research' of some forgotten doctor at Dachau." After filing the report, he handed over a sheaf of articles about the study. The following day, he later recalled, his boss called him in and said, "These people are volunteers, it says so right here. Volunteers with social incentives, it says so right here. This is all nonsense."[73]

Buxtun fumed for a year and then went higher up the chain of command, writing to the chief of the Venereal Diseases Division at the Centers for Disease Control. In March 1967 he was called to the CDC for a conference and given a condescending lecture about the importance of the study. Another few months passed. In April 1968 Martin Luther King Jr. was assassinated. Buxtun tried the CDC again, this time pointing out that a study of this kind with exclusively African-American subjects was "political dynamite." Belatedly acknowledging that the study might be a public relations problem, the CDC convened a meeting in 1969. After much discussion, it decided, on balance, to continue the study. The majority argued that treating with penicillin at such a late stage of syphilis would make no difference to the health of the subjects, whereas the data from their prospective autopsies was too useful to give up.[74]

Buxtun went on talking about Tuskegee to anyone who would listen. Finally, in 1972, he raised it at dinner with a group of friends, among whom was a reporter. She had heard him talk about it before and "glazed over," but this time she listened. Buxtun sent her a sheaf of documents and she took them to her boss, who handed the story over to Jean Heller at the Associate Press's Washington office. On July 26, 1972, it made the front page of the *New York Times*, under the headline "Syphilis Victims in U.S. Study Went Untreated for 40 Years."

The language of the article was urgent and direct: "Human beings with syphilis, who were induced to serve as guinea pigs, have gone without medical treatment for the disease, and a few have died of its late effects." The officials who had initiated the research were long retired, Heller noted, but current Public Health Service doctors were continuing the study despite "serious doubts about [its] morality." A senator was quoted as calling the research "a moral and ethical nightmare."[75] The next day the paper ran an interview with a survivor—the grandson of slaves—who recalled how the researchers recruited subjects on the pretext that they were getting free treatment. In the same article, a researcher from the Centers for Disease Control described the study as "almost like genocide."[76] On July 30, the paper's science writer noted that the study was "begun in the year Hitler came to power. It was Hitler's atrocious 'experiments' done in the name of medical science which led after World War II to the promulgation of the Nuremberg Code."[77]

The timing of the revelation, at the height of opposition to the war in Vietnam, was a disaster for the medical profession. With its explosive racial dimension, Tuskegee joined an ever-lengthening list of indictments of American hypocrisy and brutality. In 1970 Telford Taylor, chief prosecutor at the Nuremberg Trials, had published a book called *Nuremberg and Vietnam: An American Tragedy*, accusing the US government of committing war crimes in Indochina. The Kent State shootings that same year—in which four unarmed students at an antiwar demonstration were killed by members of the Ohio National Guard—had confirmed the left's worst fears about a police state. In 1971 the American Indian Movement had launched a series of protests and occupations to draw attention to the racial violence and broken treaties upon which the United States had been founded. Tuskegee brought this same sense of bitter disillusionment with the institutions of liberalism and democracy to bear on the medical profession. The cover of Taylor's book featured a swastika superimposed on the American flag, summing up the problem in a single searing image: did America have its own version of an "Auschwitz generation"?

Reporters started to show up in Macon County, Alabama, tracking down study survivors and asking them for statements. One of them, Charles Pollard, was stopped by a journalist while trading cows in a stockyard and told that he was the survivor of a sinister government experiment. Pollard was active in a local civil rights organization. Once he had pieced together the story of the study, he went to the office of Fred Gray, the legendary local attorney who had defended Rosa Parks and Martin Luther King during the heyday of the civil rights movement. Gray agreed to file a class action suit against the government for damages to the survivors and their families.

Clearly, the Department of Health, Education and Welfare had to act. A panel of nine prominent citizens, five of them African-American, was hastily convened to determine whether the study should be stopped, and whether research oversight was adequate to protect human subjects. In October 1972 the ad hoc committee urged that the men receive treatment without delay. Despite the obvious political urgency of the matter, however, the Centers for Disease Control, paralyzed by legal and bureaucratic details, seemed unable to bring the Tuskegee Study to a close by providing treatment to the survivors.

THE KENNEDY HEARINGS

Congress had been holding hearings on the same ethical issues for the last five years, but organized resistance from the research establishment and medical profession ensured that the legislative proposals fizzled out before reaching the floor of the Senate. In 1971, realizing that the issue of human experimentation was looming larger and larger in the public eye, Edward Kennedy artfully reframed the debate as "heath, science, and human rights."[78] The Tuskegee revelation proved his point that this was a civil rights issue. Seizing the opportunity created by the public response to Tuskegee, and focusing for the first time on the experiences of the *victims* of scientific research abuses, Kennedy organized his hearings on human experimentation.

On February 21, 1973, Kennedy called the first meeting to order and made some opening remarks. After listing some of the notable technical achievements of postwar medicine—heart and kidney transplants, treatments for Hodgkin's disease, a vaccine against measles, cracking the genetic code—he posed a series of rhetorical questions, culminating in what had emerged as the definitive moral dilemma for American medical research: "When may a society expose some of its members to harm in

Figure 3. Senator Edward M. Kennedy presents the Robert F. Kennedy Journalism Award for best newspaper coverage to Jean Heller of the Associated Press, in Washington, DC, April 26, 1973, in recognition of her series on the Tuskegee syphilis study. The image unites the woman who brought the Tuskegee Study to light with the man who finally put a stop to it. The award itself, a bust of Kennedy's assassinated brother by sculptor Robert Berks, reminds us of the background of political tumult and violence against which the human experimentation controversy unfolded. AP Photo/Henry Griffin.

order to seek benefits for the rest of its people?"[79] The witnesses who were called upon to answer this question were a mixture of scientists, research administrators, activists, and, most strikingly, dispossessed and marginalized American citizens who had found themselves on the sharp end of the research enterprise.

The first issue on the table was Upjohn's injectable synthetic hormone Depo Provera. With a contraceptive effect that lasted three months from a single shot, Depo Provera was the drug of choice for public health officials attempting to prevent pregnancy among people deemed unlikely to use other methods of birth control. It had not yet, however, been licensed for contraceptive use by the FDA, on the grounds that it caused cancer in beagles. The regulatory situation, in other words, was similar to that of thalidomide in the United States before its withdrawal from the market in the rest of the world. And as in the case of thalidomide, it appeared that doctors were administering the drug without informing their patients that it was an unlicensed experimental product.

Three witnesses were called to outline the issue: the lawyer and the gynecologist who together had conducted an investigation of the problem, and one of the women they had interviewed on their fact-finding mission, twenty-one-year-old Anna Burgess of Monterey, Tennessee. "We want to welcome you very much," Kennedy said to Burgess. "You are among friends, and we are looking forward to hearing from you this morning." After a damning summary of the case from the lawyer—"The informed consent of these women was not obtained before the drug was administered. None of the women were told that the FDA had not approved Depo Provera for contraceptive use"—Kennedy gently invited Burgess to testify: "You have come a long way. What would be terribly interesting, if you feel comfortable, is for you to tell us a little bit about how you came to take Depo Provera."

Burgess then opened a window onto a world very distant from Capitol Hill. From the age of five, she testified, she had suffered from a crippling bone disease. She now lived in a house in Tennessee with no running water or electricity, shared with her mother, three brothers, and her son, who was just shy of two years old. The only household income was her $99 monthly welfare check. In July 1971 she had been summoned to the welfare office, where she was asked if she was on any form of birth control. The "welfare lady . . . said I ought to be taking something or other. She said they'd rather feed one young'un as two." A few months later, Burgess was sent to a doctor, who administered a pap smear, asked her to roll over onto her side, and gave her a shot in the hip. After experiencing pain as well as severe psychological side effects, she went to a private doctor and asked for help. He had never heard of the drug she'd been given and made a phone call to the health department. He then gave her an injection to bring on her period. A week later she started to bleed: "It lasted three to four weeks. I passed about a pound of flesh."

Kennedy listened carefully to Burgess's narrative, then asked for clarification about why she had felt she had to accept the shot: "From the impression I got, if I did not take birth control, they would take the check," she said. The senator paraphrased her statement: "If you did not take the Depo Provera, they were perhaps going to threaten your assistance program?" She agreed. Later, Kennedy moved on to the question of side effects:

"You felt it made you more nervous?"

"I was so nervous, I couldn't stand nothing."

"After you took the drug?"

"After I took it. I stand around and hear my baby crying—I couldn't stand to hear the baby crying or nothing else."

"And before taking the drug?"

"I didn't have no trouble at all."

While Kennedy had been at pains in his opening statement to stress that there was "no malice"[80] involved in the utilitarian ethos of the medical profession, this testimony moved him to talk of the "outrageous treatment" of the women coerced into accepting the drug.[81] Leonard Brooks, the man in charge of the Southwest Tennessee Family Planning Project during the time Burgess received her contraceptive shot, was very upset: "I feel alone. I do not have a phalanx of bureaucrats supporting me here and I am alone." Gently invited to proceed with his testimony, he responded with accusations of paternalism, sexism, and interference on the part of the senators: "I notice that this is a group almost exclusively of middle-aged males, dictating to women how they are going to control their reproductive lives."

Brooks defended his program on feminist grounds: "Daily our women are sentenced by inadvertent pregnancy to compulsory motherhood, to loss of self-esteem, to denial of self-actualization and to instant poverty," he declared. "Where other methods have failed many have had a pregnancy free interval by use of Depo Provera. A sensitive government eases the additive and cumulative pain of so many of its people."[82] Both the accusers and the accused invoked patient self-determination to make their case, a measure of how far the debate had progressed from the time just over a decade earlier when Senator Kefauver opposed mandating informed consent on the grounds of "physicians' rights."

On the fifth day of the hearings, three ex-offenders gave testimony about research in American prisons. One of them, Leodus Jones, narrated the series of debilitating experiments that he had undertaken in order to raise money for bail. "I was told it was some kind of germ from

India," he said. "It was placed on my arm and it sat there for about five days. Then an infection set in. It began to swell. The infection lasted and, when I was released, it had not really healed."[83] He and his fellow ex-inmates submitted a written statement declaring, "These degrading conditions which require the trading of human bodies for money make any claim of *voluntary* participation by prisoners in human experimentation a cruel hoax."[84]

Called as a witness, the left-wing British journalist Jessica Mitford, author of a recent *Atlantic Monthly* article on the issue, concurred: "The drug companies . . . can buy human subjects for a fraction less than one-tenth of what they would have to pay students or other free-world volunteers. They can conduct experiments on prisoners that would not be sanctioned on student-subjects at any price because of the degree of risk and pain involved." In conclusion, she made a provocative suggestion: "Spokesmen for the drug industry claim that participation in research is a rewarding experience for prisoners, who feel that they are expiating their crimes and helping mankind. How much more rewarding, then, might it be for the stockholders in the giant drug companies to volunteer for this noble cause, since they stand to gain financially and could incidentally expiate the many crimes of the companies in which they have invested."[85]

The discussion of prison research demolished the calculus that had picked out institutionalized citizens as suitable research subjects. The very rationale that had justified it—human experimentation made only a negligible difference to lives that were already dangerous and difficult—was recast as evidence of prejudice and opportunism. "We have heard that those who have borne the principal brunt of research—whether it is drugs or even experimental surgery—have been the more disadvantaged people within our society; have been the institutionalized, the poor and minority members," Kennedy noted. "Obviously the society generally benefits from the work that is being done, the research in the biomedical field and the successes, but the risks are not generally shared."

The Berkeley criminologist to whom this remark was addressed agreed:

> Your point is tremendously well taken, that the majority of research—
> whether it be psychosurgery or experimentation with new drugs—are done
> precisely on those individuals who have least to say about their own destiny,

who are less educated, and who are least likely to have much understanding of what is going on. They tend to be dependent individuals who lack
the ability or who are not in a position, to protest and complain. . . . Their
socioeconomic situation: their lack of education and their status in many
institutions which makes them peculiarly available for this kind of research
also makes this research and these procedures ethically improper.[86]

The end had come for human experimentation justified purely on the
grounds of the greater good. "Researchers defend their work by pointing to the benefits they may turn up 'for society' when they 'pursue truth
and knowledge,'" one lawyer remarked with sarcastic emphasis, "but it is
doubtful that they . . . have the proper capability or authority to judge all
the societal costs and benefits of the means or consequences involved in
their work."[87]

If trust in medical research had reached its zenith with the successful
trials of the polio vaccine in 1954, the Kennedy hearings were the corresponding nadir. Sharecroppers, former prisoners, and welfare recipi-
ents talked back to the self-legislating institutions of the medical establishment and found a receptive ear at the highest levels of government.
Philosophically, the thesis was clear: it was the utilitarian relationship
between ends and means—the elitism of scientific reason, prediction,
and control—that had distorted the moral compass of the healing arts
and brought Cold War American medicine into such disastrous moral
proximity with its Nazi counterpart. In the tumultuous decade between
John F. Kennedy's assassination and the Watergate scandal, informed
consent went from being one of many techniques of justification and negotiation in human experimentation to an absolute precondition. With
this transformation, the assumption that medical researchers were qualified to calculate benefits and harms on behalf of others was transformed
into a species of wickedness.

UTILITARIAN NEUROLOGY

Most of the researchers who testified at the Kennedy hearings did not
even attempt to defend their actions on the grounds of the greater good.
But the scientists who gave testimony on the day devoted to the human
brain refused to admit that anything was wrong with medical utilitarian-

ism. The first witness, Bertram S. Brown, director of the National Institute of Mental Health, laid out his philosophical position at the beginning of his testimony, suggesting that it might be "immoral to withhold from human subjects the results of laboratory experiments with high potential for furtherance of scientific knowledge to 'help suffering humanity.'"[88] He was followed by Orlando Andy, a neurosurgeon from Mississippi, who opened with a series of clinical success stories, before venturing that it would be "unethical" to deny psychosurgical treatment to his patients.[89] Despite the decade-long critique of utilitarian reasoning in the law courts, the media, and the medical profession itself, these witnesses argued that it would be unethical to *restrict* human experimentation, that their highly invasive procedures on the brains of subjects should be judged by their long-term consequences, and that they had a corresponding license to self-legislate for the greater good. No other witnesses at the hearings took such a bold and unambiguously utilitarian position.

The next witness was the antipsychiatry activist Peter Breggin, the voice of the movement for informed consent as applied to the behavioral sciences. In 1970 Breggin had set up an organization called the Project to Examine Psychiatric Technology and embarked on a crusade against behavior-modification techniques, especially psychosurgery and electrical stimulation of the brain. A barrage of articles in both the popular press and medical journals resulted in some significant political victories, and by 1973 he had an impressive track record of blocking funds for this type of neuroscientific research. At the hearings, he denounced the "mechanistic, anti-individual, anti-spiritual view," which "gives justification to the mutilation of the brain and the mind, in the interests of controlling the individual." Researchers enjoyed "an elitist power over human mind and spirit," he claimed. "If America ever falls to totalitarianism, the dictator will be a behavioral scientist and the secret police will be armed with lobotomy and psychosurgery."[90]

Next to testify was a scientist whose work seemed to exemplify the problems Breggin was trying to articulate: Robert Galbraith Heath, chairman of the Department of Psychiatry and Neurology at Tulane University. Slim, dapper, and silver-haired, the very picture of a gentleman scientist, Heath described how his laboratory was dedicated to the treatment of nervous system disorders using electrodes implanted into "precise, predetermined brain regions." Since 1950, he testified, sixty-five patients had undergone the procedure in his laboratory at Tulane. The electrodes "remain securely in the brain sites for many months," he explained, allowing him "to build a meaningful bridge between mental activity and physical activity of the brain, the organ of behavior."

According to Heath's testimony, what connected mental states, brain activity, and behavior were circuits of pleasure and pain: "With our techniques, we have demonstrated brain sites and pathways which are involved with pleasurable emotional states and those involved with painful emotional states—such as rage and fear, which are basic to violence and aggression." He claimed, in short, that he could control his subjects' moods and emotions by flipping a switch: "Stimulation of brain pleasure sites induces feelings of well-being, whereas stimulation to adversive sites induces violence, rage and related adversive emotional states."[91]

Heath reassured Kennedy that he and his team "insist that all patients as well as family members are thoroughly informed of the nature of the procedures and of every possible complication before forms are signed." But then he shifted the focus of the discussion:

> There are also ethical questions from another viewpoint. Is a doctor justified in withholding treatment with a procedure that might carry some risk if the alternative is the certainty that the patient will spend the rest of his life in the back wards of a mental institution? . . . any procedure which offers even minimal hope of preventing chronic hospitalization—even if it is risky—is worthy of trial since life in the back wards of a chronic mental hospital is worse than death.[92]

According to Heath's utility calculus, in other words, the horrible conditions on psychiatric wards justified even his riskiest interventions. This was the ultimate archangelic reckoning. Because he judged life in a mental hospital to be worse than death, he was licensed to do just about anything to a psychiatric patient.

It had long been the practice of the Tulane laboratory to film the stimulation experiments, and as part of his testimony Heath projected three short movies, explaining that they showed his patients undergoing stimulation of their pain and pleasure centers. The first film showed a patient "in whom we turned on the adversive brain circuitry to induce violent impulses." As it began to roll, Heath explained to Kennedy what they were watching: "This is the start of the stimulation. The patient begins to cry out with rage. He tells the doctor who is nearest to him, a neurosurgeon, that he is going to kill him; that he wants to murder him. As soon as the stimulus goes off, this emotional state disappears." The second and third films, Heath explained to Kennedy, showed the patients' "pleasure sites" being stimulated.[93]

I have not been permitted to view any of the Tulane footage, but we do have a description of the first film played for Kennedy, written by two journalists who were shown it by Heath himself in the 1990s:

> The next film shows a patient having his "aversive system" stimulated. His face twists suddenly into a terrible grimace. One eye turns out and his features contort as though in the spasm of a horrible science fiction metamorphosis. "It's knocking me out. . . . I just want to claw . . ." he says, gasping like a tortured beast. "I'll kill you. . . . I'll kill you, Dr. Lawrence."

Another passage from the same source conveys a vivid impression of one of the pleasure experiment films:

> A woman of indeterminate age lies in a narrow cot, a giant bandage covering her skull.
>
> At the start of the film she seems locked inside some private vortex of despair. Her face is as blank as her white hospital gown and her voice is a remote, tired monotone.
>
> "Sixty pulses," says a disembodied voice. It belongs to the technician in the next room, who is sending a current to the electrode inside the woman's head. The patient, inside her soundproof cubicle, does not hear him.
>
> Suddenly she smiles. "Why are you smiling?" asks Dr. Heath, sitting by her bedside.
>
> "I don't know. . . . Are you doing something to me? [Giggles] I don't usually sit around and laugh at nothing. I must be laughing at something."
>
> "One hundred forty" says the disembodied technician.
>
> The patient giggles again, transformed from a stone-faced zombie into a little girl with a secret joke. "What in the hell are you doing?" she asks. "You must be hitting some goody place."[94]

After watching the films in silence, Kennedy continued with his questioning: "What you are really talking about is controlling behavior," he suggested. Heath bridled at this: "I am a physician and I practice the healing art. I am interested in treating sick behavior—not in controlling behavior." Undaunted, Kennedy went on: "You have shown and testified about how you can replicate pain and pleasure by the implantation of these electrodes in different parts of the brain." This time Heath assented. Kennedy continued: "This is behavioral control. As I understand it, you are trying to use this technique to treat people." Again, Heath agreed. Kennedy then asked, "Would it not be adaptable to treat other people as well, normal people?" to which Heath replied: "I think it would be, but I think normal adaptive people are already being treated. I am sure Dr. Skinner is going to talk on that. Our learning experiences, our attitudes are modified every day."[95]

After being reassured that Heath's technique was too intricate and expensive to be applied to more than a handful of people each year, Kennedy dismissed the neuroscientist with thanks. "Our next witness needs very little introduction," he announced. "Dr. Skinner, you may proceed with your statement, sir." At this prompt, B. F. Skinner leaned in to the microphone and launched into a prepared statement. There was no need for technological or chemical intervention, he claimed, because "behavior is selected and strengthened by its consequences—by what the layman calls rewards and punishments, and this fact has long been exploited for purposes of control." Noting that pain and punishment had historically been the main source of insights about human conditioning, Skinner argued that the time had come to switch over to pleasure and reward, which were both more humane and more effective: "Under punitive control a person does not feel free."[96]

Freedom, for Skinner, was just a misleading name for the pleasant sensation that people experienced when they could get their hands on what they wanted. The idea that liberty or autonomy might represent anything more precious or profound was ludicrous to him. "The prisoner of war who resists efforts to demean him or change his views is not demonstrating his own autonomy," he scoffed, "he is showing the effects of earlier environments—possibly his religious or ethical education, or training in techniques of resistance in the armed services."[97] In dismissing the whole concept of autonomy, Skinner revealed his contempt for the

ideal of the free, sovereign individual that underlay the movement for informed consent. Psychological autonomy was nothing but a prescientific
superstition. Freedom was a grand but empty word. The only things that
truly moved man were reward and punishment. With self-determination
ruled out of court, it followed that all that mattered for ethics was the
consequence of any course of action for the balance of pleasure and pain.

It is not difficult to see the archangelic appeal of this worldview.
Heath and Skinner were scientists, dedicated to the goal of prediction
and control. By manipulating reward and punishment, they believed they
had grabbed hold of the two great levers that directed the machinery of
human behavior. They were philanthropists, devoted to the ideal of the
amelioration of suffering. By working directly on pain and pleasure, they
felt they had tapped into the purest source of good works. They were
truth seekers, committed to stripping away superstition. In their astrin-
gent analysis of human behavior, they tasted the clean flavor of reality
unsweetened by the illusion of freedom.

Other American researchers had discovered that scientists were not,
in fact, archangels. They had been forced by the pressure of public opinion to acknowledge that the greater good could not be grasped directly. It
had to route itself through painstaking communication and negotiation
with other humans, assumed to be free. For the behavioral scientists,
however, claims of psychological autonomy flouted everything they believed about the nature of the human nervous system. The idea that the
human animal was motivated only by pain and pleasure was such a powerful commitment on their part that they found it impossible to admit
moral freedom into their research ethics. And without a commitment to
moral freedom, the philosophy of informed consent was empty.

It was a genuine revolutionary fervor that sought to topple the doctor-
as-father from his place at the head of the medical profession. Opposition
to his white-coated, paternalistic authority was embodied in the ideal of
an existentially free individual, asserting her sacred right to bodily autonomy against the unfeeling calculus of utilitarian science. The confrontation between the two sides of the human experimentation debate brought
the first principles of political ethics—utility versus autonomy, the greater
good versus human rights—into a fierce dispute about the moral, medical, psychological, and neurological constitution of humanity. As the rest
of this history will show, the liberties and sufferings of research subjects

were but acute manifestations of the condition of liberal subjects in general. The Kennedy hearings, trafficking in the stuff of life and death, pain and pain relief, laid bare the starkest outlines of a philosophical conflict that was centuries old. We will catch up again with Heath and Skinner and their adversaries in the penultimate chapter, but for now we will travel back to the deep historical origins of the clash between patients' rights and medical utility. In 1789 Jeremy Bentham enthroned his "two sovereign masters, *pain* and *pleasure*," as the ultimate arbiters of right and wrong. Against the assertion of rights and liberties on the part of the French revolutionaries, he proposed the test of utility as a less perilous path to social progress. Scientific medicine subsequently became the domain in which his utilitarianism found its purest expression. When patient autonomy triumphed over medical utility in the last months of the Nixon presidency, in other words, the Rights of Man won a belated victory over their most influential nemesis.

2

Epicurus at the Scaffold

THE AUTO ICON

Despite London's policy of making motorists pay for the privilege of driving in the city center, Gower Street in Bloomsbury still thunders with traffic all day and most of the night. Entering through the main gate of University College on a summer day affords little relief from the urban roar: the main quadrangle is as democratic as a city park. Office workers lounge about on the steps of the Greek Revival main building, smoking cigarettes and eating lunch. Homeless people snooze in the shade of dusty trees. Litter somersaults merrily over balding grass.

Turn right and head for the south cloisters and you will find one of London's most eccentric memorials. In a large, glass-fronted mahogany box, like a telephone booth in an old-fashioned hotel, sits Jeremy Bentham's skeleton, padded and dressed in his clothes, topped with a wax head. He wears a blue frock coat, britches, white stockings, and black buckled shoes. One plump, gloved fist rests on a beloved walking stick. The head has a benign, grandmotherly air: the hair is long, wavy and blond; the blue glass eyes twinkle in the jowly moon face; the tall straw hat is set back at a jaunty angle. The effigy radiates immortal purposefulness, the opposite of eternal repose. This is, after all, the man who coined

Figure 4. Jeremy Bentham's Auto Icon. This photo, taken in 1948 to accompany a scholarly article, includes Bentham's preserved actual head, staring at the viewer with zombie astonishment from between the master's feet. The head is now kept in temperature-controlled storage, but the rest of the display can still be visited in the South Cloisters of University College London. Photo by C. F. A. Marmoy, courtesy of Alpha Photo Press.

the phrases "false consciousness" and "sexual desire," the lawyer whose penal codes were adapted for use from Delhi to Louisiana, the philosopher of surveillance who updated ethics for the industrial age.

Bentham was only twenty-one when he made a will leaving his body to science. The year was 1769, and he may have been the first person ever to make this gesture. By the time of his death in 1832, plans for his posthumous existence had become as grand, bizarre, and intricate as the man himself. After consulting with his physician, and reading up on the mummification techniques of native New Zealanders, he left elaborate instructions for the public dissection of his body, followed by the preservation of his effigy in a glass box. He baptized this shrine the "Auto Icon" and fondly imagined whole galleries of such monuments to enlightened men.[1]

By leaving his corpse to be dissected, Bentham hoped to contribute to the science of anatomy. By instructing that his body then be reassembled into an icon, he left an enduring memorial to the gesture itself, encouraging future generations to aid the cause of medical progress. In this way, he hoped to wring every last drop of medical utility out of his own dead body. As the only explicit monument to medical utilitarianism ever erected, the Auto Icon stands as the symbolic fountainhead of the human experimentation scandals of the 1960s and 1970s.

The previous chapter explored the aspect of medical utilitarianism that owes the most to Bentham's legacy: the bitter opposition between individual rights and the greater good. When Bentham celebrated medical utility, he did so in explicit opposition to the notion of inalienable rights. His aversion to the concept arose early in life. As an intellectually precocious Tory teenager, he dismissed the appeals to "natural rights" in Whig legal theory on the grounds that they were hopelessly vague. As a loyal monarchist during the American Revolution, he deplored the hypocrisy of murderous colonial rebels proclaiming a "right to life, liberty and the pursuit of happiness." A few years into the French Revolution, he memorably denounced the Rights of Man as "nonsense upon stilts." In contrast to the vapid rhetoric of rights, the balance of pleasure and pain was, for Bentham, measurable, quantifiable—*real*.

Bentham's steady focus on brute sensation forged a direct link between politics and medicine. Bad government produced unhappiness. Retributive justice caused unnecessary pain. The human nervous system was where the badness of bad laws appeared as physical and mental suffering. The political, for Bentham, was physiological. This meant that legal reform was civil medicine—in the most literal sense. Conversely, scientific medicine was the purest and most concentrated form of civil good. In Bentham's hands, medical progress and utilitarian reform were welded into a single improving force by virtue of their shared concentration on pain and its relief.

Bentham did much to advance these ideas, but utilitarianism did not leap full-formed from his brow. A tradition of thinking about politics in terms of pain and pleasure went back to the English Civil War of the seventeenth century. After the execution of Charles I, a member of the court in exile in Paris, Thomas Hobbes, then sixty years old, started composing a treatise on politics. One of Hobbes's closest friends in Paris was a devotee of the ancient philosopher Epicurus, who held that pain was the only evil and pleasure the only good. The stark truth of this creed struck Hobbes with overwhelming force as he saw the devastation caused by the English Civil War. In 1651, in the aftermath of the regicide, he published his *Leviathan*, an infamous work of Epicurean political philosophy that linked pain and pleasure to political obedience. On the ground of pain aversion, Hobbes argued, safety and security must trump rights and liberty. Because revolution *hurts*, he announced, absolute submission to the sovereign power is the overriding duty of any rational sentient being.

Unlike Bentham, Hobbes believed in the reality of natural rights. That did not mean he liked them. The rights and liberties claimed by Parliamentarians, regicides, Diggers, Levellers, Anabaptists, and other revolutionary fanatics were natural, Hobbes conceded, but also dangerously anarchic. In a state of nature, it was the right of every man to take what he could get. But because we humans hate pain and fear death, we are forced to form a civil society to secure our bodily safety.

Over the course of the eighteenth century, British pleasure-pain psychology swiveled 180 degrees away from Hobbesian absolutism. By the 1750s it had been completely reconciled with Enlightenment ideas about rights and freedoms. But then came the American and French Revolutions. Violent political upheaval raised the specter of the English Civil War, making Hobbes's cynicism seem freshly relevant. It was the regicide of the French king that sealed the deal. Alone among the European nations, the British viewed the French Revolution from the perspective of their own doomed attempt to execute a king and establish a republic. Lessons had been learned about the costs and benefits of violent insurrection.

Thomas Hobbes appealed to pain and pleasure—the barest fundamentals of our animal existence—to cram a head back on the decapitated body politic; Jeremy Bentham did the same to avert the threat of another English regicide. It is a set of mirror images—of the English Civil War and the French Revolution, of Charles I and Louis XVI, of Hobbes and Bentham—that explains why utilitarianism is British to the core. On the scaffolds where kings were killed, Epicurus was updated for the purposes of counterrevolution.[2]

HOBBES THE EPICUREAN

So, what kind of a man was this Thomas Hobbes, who brought pain and pleasure into politics in ways that reverberate to this day? He tells us that he was born afraid, arriving in the world prematurely on April 5, 1588, when his mother heard that the Spanish Armada was heading for English shores and gave birth to "twins, myself and fear."[3] The father of this sensitive little creature was an uneducated country parson, an alcoholic and compulsive gambler who got into a fistfight with another vicar in the graveyard of his church and fled to London, leaving his wife and children to fend for themselves. Most of what we know about the derelict parson comes from John Aubrey, Hobbes's friend and biographer, who has some fun at the senior Hobbes's expense, recalling the day he dozed off in church and shouted out "Clubs are trumps!" in his sleep.[4]

Luckily for the family, old Parson Hobbes had a brother who was both prosperous and childless. A successful manufacturer of gloves, he paid for the children's education. At school, Hobbes learned classics to such good effect that by the age of fourteen he had translated Euripides from Greek into Latin. The following year, again with financial assistance from his uncle, he entered Magdalen Hall Oxford. According to Aubrey, Hobbes frittered away his undergraduate days baiting jackdaws and gaping at maps in bookbinders' shops. He must not have disgraced himself utterly, however, because in 1608, when he reached the age of twenty, the principal of the college gave his name to William Cavendish, an immensely wealthy aristocrat who was sniffing around Oxford for a tutor for his second son.[5]

And so it was that Hobbes became a "client" of the Cavendish family, a career that lasted, on and off, for the rest of his life. William Cavendish junior—a "waster," in Aubrey's blunt estimation—was only two years younger than Hobbes. At first, Hobbes's position in the household was

more as a friend and servant than any sort of tutor: "He was his lord-
ship's page, and rode a hunting and hawking with him, and kept his privy
purse."[6] The wastrel young nobleman—who would die at the age of forty-
three, reputedly from overindulgence—often sent Hobbes to borrow
money on his behalf, and so our hero spent a lot of time standing around
in wet shoes waiting on various creditors. Aubrey makes him seem a bit
pathetic at this juncture, noting that he caught many colds as a result of
this humiliating task, but in truth young Hobbes could not have landed
a cushier gig.[7] The Cavendishes were the grandest of the grand, and for
many years the philosopher-to-be resided at Chatsworth, the gorgeous,
126-room family seat, where a whole library was eventually stocked with
books of his choosing.

Hobbes's qualifications for this position were personal ones: the
Cavendishes "lov'd his company for his pleasant facetiousness and good-
nature."[8] All extant portraits of him show a sharp-featured old man, but
in his youth he seems to have been very good-looking. Possessed of great
reservoirs of charm and cheerfulness, he was six feet tall with coal-black
hair, a reddish mustache turned up at the ends, and a little soul patch
beneath his lower lip. He was "much addicted to musique" and skillful at
the bass viol, an instrument similar to the cello.[9] Most striking were his
hazel eyes: "He had a good eie, and that of a hazell colour, which was full
of life and spirit, even to the last. When he was earnest in discourse, there
shone (as it were) a bright live-coale within it."[10]

Hobbes did not become truly intellectually serious until his early
middle age, but he was always known for being quick on the uptake.
Francis Bacon—lord chancellor and founder of the philosophy of
science—liked to stroll around the grounds of Chatsworth declaiming
his great thoughts, while Hobbes trotted alongside scribbling down his
pearls of wisdom. Bacon judged Hobbes the only one of his young aman-
uenses who actually understood what he was saying. On Continental
tours with his aristocratic charges, Hobbes rubbed shoulders with Gali-
leo and Descartes. The superstar playwright Ben Jonson was one of his
best friends. He weathered this life among the great intellects of his day
with almost pathological self-confidence, boasting that he read little of
other men's work, preferring to come to his own conclusions on the big
questions: "He was wont to say that if he had read as much as other men,
he should have knowne no more than other men."[11]

Born into poverty, obscurity, and family dishonor, Hobbes had found himself a luxurious berth with one of the wealthiest, best-connected families in the land. He was correspondingly loyal to the interests and privileges of the English aristocracy. In 1640, as Parliament began to challenge the king's sovereign authority, he composed a short treatise on the necessity for absolutist government to act as an antidote to the ineradicable unpleasantness of human nature. The book was not published, but it circulated in manuscript form and gave him the reputation of a die-hard Royalist. When Parliament proved itself to be in a fighting mood, Hobbes decided that the political temperature was getting too high for his particular combination of aggressive views and personal timidity, and took himself off to Paris. He would later boast of being the "first of all that fled."[12]

In Paris, Hobbes mixed in brilliant intellectual circles that he had cultivated on earlier Continental tours. He debated metaphysics with Descartes, judging, with his usual blithe confidence, that the great man's "head did not lye for philosophy."[13] He composed a treatise on optics. His closest friendship was with the freethinking priest Pierre Gassendi, who for decades had been working on a commentary on the writings of the ancient Greek philosopher Epicurus. Under Gassendi's influence, Hobbes became a fellow devotee of the pagan philosopher of pleasure.[14]

Epicurean philosophy is best known through the 7,000-line Latin poem *De Rerum Natura*, "On the Nature of Things," by the Roman poet Titus Carus Lucretius. Dedicated to Venus, the poem is a hymn to fleeting happiness and earthly beauty, abounding with flower garlands, piping shepherds, wine, birdsong, and honey, and including some vivid descriptions of sexual passion. The work appeals to the universal human experience of nature to mount a series of speculations about the structure of the physical world and the meaning of freedom. It ends on a shattering existential note, with a vividly repellent description of the Plague of Athens. The emotional extremes of the poem—from the exquisite sensations of romantic love to the searing agonies of a devastating epidemic—define the ethical duality of Epicurean philosophy: for Epicurus and his followers, the greatest evil in the world was pain, and pleasure was the greatest good.

Following Epicurean morality to its logical conclusion, Hobbes's friend Gassendi once tentatively remarked, "I suppose that which is

Profitable and that which is Good to be but one and same thing."[15] But the priest was too sincerely devout to stay with such a materialist sentiment. Drawn to Epicureanism because it offered a promising framework for scientific inquiry, he also insisted that it was compatible with the omnipotence of God and the immortality of the soul. Some of his fellow scholars went along with this implausible compromise. Not Hobbes. Of all the Epicureans of the time, Hobbes alone was ready to follow the moral primacy of pain and pleasure to its logical end and apply it to the problems of the day. Transposed to an age of political crisis, an ancient pagan hymn to pleasure was reworked by Hobbes into a credo of deference to power.

In 1642, as Parliament and king went to war, Hobbes published a small edition of a political tract mounting an Epicurean argument for peace at any price. For the sake of bodily safety and integrity, he argued, we must give up most of the rights that we enjoy in a state of nature and defer to a sovereign power capable of imposing order. From Paris, he watched with dismay as the rebel forces turned the tide of war in their favor. In 1645 Charles slipped away to Scotland. At the same time his son, the Prince of Wales, later Charles II, went into exile in Paris, where Hobbes became, for a short season, his mathematical tutor.[16] As the situation worsened for Royalists, the demand grew for a new edition of Hobbes's defense of monarchy, which came out in 1647 under the title *De Cive* and began to garner him an international reputation.

De Cive, "The Citizen," contains much of the essence of Hobbes's political philosophy, including the insight—so important for the utilitarian legal philosophers of a later century—that criminal punishment should serve no ends other than deterrence: "We must have our eye not at the evil past but at the future good . . . that the offender may be corrected, or that others warned by his punishment may become better."[17] The Epicurean focus on worldly happiness had led Hobbes to the ultimate utilitarian conclusion: the only thing that counts for morality is the *consequence* of our actions. Nothing else is important. We should not concern ourselves with motivation, or purity of heart, or retributive justice, or obedience to unbreakable rules. All that matters is the effect we have in the world.

In 1649 Hobbes's fears were realized when Charles I was arrested on charges of treason, tyranny, and murder and sentenced to death. On a bitterly cold day at the end of January, the king of England was led to a

scaffold at Whitehall. A huge crowd had assembled to witness the culminating symbolic act of the revolution. "I am the martyr of the people,"
Charles announced, before tucking his own long hair into his cap and
laying his neck on the block. After a few moments of silence, he issued his
final command as sovereign and lawgiver: "The King stretching forth his
hands, The Executioner at one blow, severed his head from his Body and
held it up and shewed it to the people."[18] At the sight of the sovereign's
dripping head, a great groan of lamentation rose up from the crowd.

Hobbes began to write his most famous book in response to the regicide.[19] Composed with preternatural speed, *Leviathan; or, The Matter,
Form and Power of the Commonwealth, Ecclesiastical and Civil,* is one
of the most enduring and influential philosophical treatises ever written
in the English language. The first edition, published in London in 1651,
features a title page on which a "body politic," made up of legions of tiny
men and topped with a crowned head, looms over a walled city and fields.
The figure holds a sword in his right hand and a bishop's crosier in his
left. On either side of the book's title are arrayed the symbols of the sovereign's worldly and spiritual power. The image neatly sums up the book's
argument: once individuals have transferred power to the sovereign for
the sake of their own protection, they lay down their right to rebellion
against his sole and absolute authority.

For Hobbes, repairing the damage wrought by the Civil War would
require nothing less than a science of mankind based on mechanical first
principles. Appealing to the bedrock experience of pain and pleasure, he
made his argument for peace at any price. Appetite and aversion, Hobbes
claimed, defined good and evil: "Whatsoever is the object of any mans
Appetite or Desire; that is it, which he for his part calleth *Good*: and
the object of his Hate, and Aversion, *Evill.*" All morality was relative to
human impulses and drives: nothing "to be taken from the nature of the
objects themselves" determined their moral content.

To illuminate this somewhat cynical view of human morality, Hobbes
clustered "Beautifull," "Handsome," "Gallant," "Honorable," "Comely,"
and "Amiable" under the heading of good. For their opposites he offered
"Foule," "Deformed," "Ugly," "Base," and "Nauseous." The former were
"Utile, Profitable;" the latter "Inutile, Unprofitable, Hurtfull."[20] Here
we see the first stirrings of utilitarian psychology. Morality is reduced to

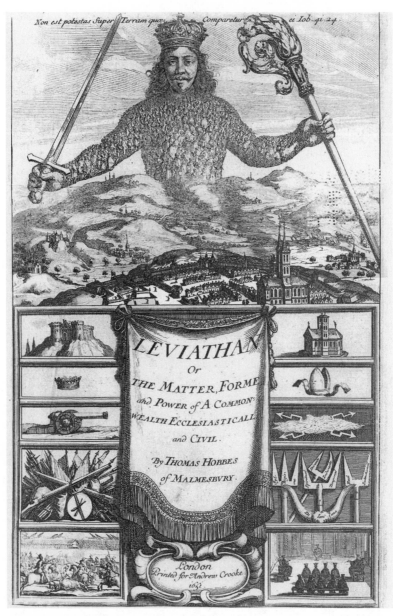

Figure 5. The frontispiece of Thomas Hobbes's *Leviathan*. Hobbes argued that absolutist government was the only way to secure the safety of the people, symbolized here by the legions of tiny figures that make up the body politic. In reinventing absolutism in terms of bodily security, Hobbes tied state power to the imperatives of pain and pleasure. These arguments returned with renewed force at the time of the American and French Revolutions, when Jeremy Bentham argued against the Rights of Man and for the test of utility.

sensation. Goodness is nothing more than the pleasure and desire associated with utility, profit, and beauty; badness no more than the pain and disgust engendered by loss, deformity, and sickness.

Against the overriding imperatives of peace, Hobbes ranged the seductions of political freedom. These, he urged, must be resisted at all cost. Our most primordial impulses entail, as a matter of necessity, that each one of us "give up my Right of Governing my selfe" or risk conflict, premature death, and all the other brutalities of a state of nature.[21] This is the choice before us, according to Hobbes: take the "Utile" path of deference and obedience, or endure the perils and horrors of self-government. Not only does *Leviathan* outline a nascent form of utilitarian psychology, it also lays out the opposition between utility and autonomy (the latter— from *auto-nomos*, or self-legislating—being a fair translation of Hobbes's "Right of Governing my selfe").

Much of the persuasive power of Hobbes's tract lies in his comic cynicism about human nature. The Hobbesian homunculus is vain, acquisitive, and competitive, insatiable in his pursuit of knowledge, fame, wealth, and power. Life is a race, and he is built to run it. He delights in lording it over his fellows, and his sense of humor belongs squarely in the monkey cage: "*Sudden Glory*, is the passion which maketh those *Grimaces* called LAUGHTER; and is caused either by some sudden act of their own, that pleaseth them; or by the apprehension of some deformed thing in another, by comparison whereof they suddenly applaud themselves."[22] We laugh, in other words, and thump our hairy little chests, when we witness the humiliation of others. By nailing the psychology of the comic pratfall, Hobbes rubs the reader's nose in her drive to self-glorification at the expense of her fellows. This ability to generate an uncomfortable sense of recognition is what makes his conclusions so hard to dismiss, even at the distance of three and a half centuries.

The most famous four words in the book—"nasty, brutish, and short"—occur at the end of a sentence describing life in a state of nature:

> In such condition, there is no place for Industry; because the fruit thereof is uncertain: and consequently no Culture of the Earth; no Navigation, nor use of the commodities that may be imported by Sea; no commodious Building; no instruments of moving, and removing such things as require much force; no Knowledge of the face of the Earth; no account of Time;

no Arts; no Letters; no Society; and, which is worst of all, continuall feare, and danger of violent death; And the life of man, solitary, poore, nasty, brutish, and short.[23]

The chapter in which the celebrated sound bite appears is entitled "Of the Naturall Condition of Mankind, as concerning their Felicity, and Misery." The Epicurean twins Felicity and Misery would go on to have glorious careers in the country of Hobbes's birth. Pleasure's commands and calculations evolved by degrees into Benthamism and Malthusianism, while the miseries of the "primitive" state became an important pretext for British imperial expansion. *Leviathan's* pleasure-pain psychology would go on to underpin theories of political economy, evolutionary biology, and behavioral science. With Hobbes's famous book, Epicurean psychology and ethics arrived in the modern world—as a rebuke to regicide.

In its own time, *Leviathan* simply succeeded in enraging everybody. People across the political spectrum were offended by it, including the English court in exile. Charles I's Catholic widow was scandalized by Hobbes's attacks on the church and barred him from court. The clergy in Paris attempted to have him arrested. Exiled Royalists thought his theory of sovereignty gave succor to all the wretched turncoats in England who had made peace with Cromwell. Hobbes's basic principle, after all, was that the people owed their allegiance to the government that protected them, whatever its form, and in 1651 that form happened to be the Commonwealth of England, without a king or a House of Lords. With English monarchists and French clerics baying at his heels, Hobbes left Paris in 1652 and made his way back to London. One envious contemporary observed that he was "much caressed" by the new regime.[24] He promptly retired to scholarly seclusion and devoted the remaining years of the Interregnum to producing another two volumes of political science.

In 1658 Cromwell died. His son Richard was thrust into his place and proved unequal to the task. After only a year, the army banished Richard to the countryside, leaving England in a state of anarchy. One practical-minded army leader led his troops to London to restore order and invited the late king's son to return. In May 1660, on the day of his thirtieth birthday, Charles II entered London. He was greeted with an outburst of pagan jollity. Flowers were strewn in the path of his carriage. Balconies along the route were hung with tapestries and filled with laughing ladies. Bells rang out. The day was declared a national holiday. Maypoles were thrown up and dánced around. The fountains of Oxford spouted wine and beer.

Hobbes might have had reason to fear this development. After all, his philosophy of obedience had been embraced by Cromwell and his followers. As it turned out, he was able once again to charm his way through a reversal of political loyalty. John Aubrey tells us of the happy reunion between the new king and the renegade philosopher. Three days after his triumphant return, Charles was riding in his carriage, saw Hobbes in the street, raised his hat, and asked after his health. Thereafter the king—always amused by his puckish former mathematics tutor—gave Hobbes free range at court and a royal pension (albeit not always paid).[25]

Even in our godless day, Hobbes's *Leviathan* still packs an atheistic punch. In his own lifetime the book was considered absolutely scandalous. For many decades, "Hobbist" was a synonym for infidel, and a grave insult. On at least one occasion Hobbism was elevated to a punishable heresy.[26] In the end, however, Hobbes triumphed. His psychological and political insights outlived the passionate denunciations of his materialism and became almost axiomatic. By basing his theory of governance on the most elementary psychological fundamentals, he stripped political philosophy of sentiment, superstition, utopianism, and obscurity. And along the way, he invented the prototype of the utilitarian self, in whom reason was little more than a physiological reflex and freedom just the unopposed satisfaction of desire.

Most important for the history of utilitarian psychology, Hobbes argued for moral continuity between humans and other animals. In his relentless pursuit of pleasure and flight from pain, Hobbesian Man is little different from his brutish brethren. The "Succession of Appetites, Aversions, Hopes and Fears, is no lesse in other living Creatures then in Man: and therefore Beasts also Deliberate."[27] In later chapters we will see how this reduction of human reason to a bestial succession of appetites and aversions continued to reverberate in Anglophone medicine, neurology, economics, and evolutionary psychology, supplying a naturalistic explanation for human behavior and a series of biological templates for the reorganization of society.

RIGHTS AND LIBERTIES

Over the course of the eighteenth century, as the British body politic weathered a series of new crises without actually losing its head, Hobbesian psychology underwent a number of revisions, elaborations, and amendments. The emphasis on pain and pleasure held steady, but the absolutism softened. By the 1750s the ethics and psychology of pain and pleasure were fully reconciled to the ideal of constitutional monarchy, the separation of powers, and the Rights of Man. They became, in a word, liberal.

The transformation of British pleasure-pain psychology from absolutism to liberalism can be followed in the life of a single individual, the philosopher John Locke. Born in 1632 to a Puritan family, Locke was a sixteen-year-old pupil at Westminster School, "in earshot of the awe-stricken crowd" when Charles I was executed.[28] In 1652 he went to Christchurch College Oxford, where he was treated to the same hopelessly old-fashioned curriculum as Hobbes had been. Like his predecessor, he seems to have frittered away his student years, reading romances and cultivating the company of witty men. But he acquired a taste for philosophy from reading Descartes, and after dabbling with the church, medicine, and the law, he accepted a series of college appointments, becoming a college tutor in 1661.

Locke's family was Parliamentarian, but the events of the Civil War had soured him on revolutionary zeal. In 1660, the year of Charles II's return, he wrote an essay recapitulating in less cynical language the absolutist arguments of Hobbes's *Leviathan*. He opened on an autobiographical note: "I no sooner perceived myself in the world," he remarked, "but I found myself in a storm." Having survived this terrible political weather, his sheer relief at the advent of peace and security obliged him "in duty and gratitude" to do all he could to school his compatriots in "obedience." For Locke, as for Hobbes, no price was too high for peace and the rule of law.[29]

Soon after penning this plea for safety, quiet, and obedience in all matters, Locke set up a laboratory in his rooms, boiling, burning, and fractionating substances. He became an expert on plants, earning himself a place in the Royal Society for the Promotion of Natural Knowledge, the first scientific association in England, inaugurated by Charles II in 1660. A stint as secretary on a diplomatic mission to Brandenburg exposed him to a community where Lutherans and Calvinists lived side by side without open strife, and helped change his mind about the feasibility of religious freedom (at least among Protestant sects).[30] In 1667 he left Oxford and went to London, under the patronage of the lord chancellor, the future Earl of Shaftesbury. In London, Locke met Thomas Sydenham, one of the first physicians to group ailments based on symptoms, and collaborated with him on a series of medical treatises. A few months later, the two men were part of the medical team that recommended an operation to treat Shaftesbury's liver abscess. The surgeon, under Sydenham and Locke's direction, secured a little silver pipe through the abdominal wall in order to drain the pus. Crediting Locke with saving his life, Shaftesbury wore the pipe for the rest of his days.

Meanwhile, Charles II was gradually descending from adored guardian of peace to despised crypto-Catholic despot. Shaftesbury became the center of opposition to the king's authority, and Locke began to organize his thoughts on political resistance. His full transformation from defender of authority to champion of freedom is marked by his 1667 *Essay on Toleration*, in which he argued (in direct opposition to his earlier claims) that liberty of conscience must be granted on the grounds that the stakes were no less than the "infinite happinesse or infinite misery" of the life hereafter.[31] By transferring Hobbes's pain and pleasure psychology from earth to heaven, Locke opened up a space for political liberty. For Hobbes, the freedom to accept or reject the sovereign's leadership was nothing more than a recipe for anarchy and bloodshed in this world. Against Hobbes's pragmatic insistence on total obedience, Locke argued that the immortal soul must be granted some wiggle room, even at the price of political instability.

In 1679 Locke was summoned by his patron to work on a philosophical justification for the exclusion of Charles II's Catholic brother James from the succession. The resulting *Two Treatises of Government* is one of the great classics of liberalism. In it Locke asserted that the "Body Politik under one Government" is formed by a group of men for the "mutual Preservation of their Lives, Liberties and Estates."[32] So far, so Hobbesian, but for Locke the power of the government extended only so far as it promoted the public good, and no further. When the sovereign fails in that mission, he argued, the people "have a Right to resume their original Liberty."[33]

By the time Locke got around to publishing this subversive doctrine, the transfer of power from one sovereign to another at the behest of Parliament was an established fact. James II had, indeed, overstepped the boundaries of Locke's putative compact with the people and had been replaced—without war breaking out—by his daughter Mary and her husband, the impeccably Protestant William of Orange. This so-called Glorious Revolution achieved the goal of limiting the power of the monarchy without the traumatic dismemberment of the body politic. In 1689 William and Mary signed "An Act Declaring the Rights and Liberties of the Subject," said to have been inspired by Locke's philosophy of orderly rebellion.[34]

In 1690 Locke published *An Essay Concerning Human Understanding,* in which he laid out his famous doctrine that the human mind is a blank slate at birth and that all our ideas are received by means of the senses during our lifetimes. There is, however, an innate structure to Locke's senses, from which all morality and politics flow:

> Nature, I confess, has put into man a desire of happiness, and an aversion to misery: these indeed are innate practical principles, which (as practical principles ought) do continue constantly to operate and influence all our actions, without ceasing: these may be observed in all persons and all ages, steady and universal.[35]

For Locke, pain and pleasure—or, as he preferred to call them, "delight" and "uneasiness"—had been placed in the human frame by God in order to preserve the body from harm.[36] He agreed with Hobbes that these were primordial and overriding motives and that things "are good or evil only in reference to pleasure and pain,"[37] but there was more space within his scheme for psychic freedom. Even though we are driven to pursue delight and avoid uneasiness, we are nonetheless able to take into account the long-term consequences of our tastes and inclinations and to "suspend the prosecution of this or that desire." The most important long-term consequence was, of course, our eternal fate in the afterlife.

The ability to stop, reflect, and consider whether the pursuit of an appetite was in our best interests was, for Locke, the "source of all liberty."[38] His Anglican-liberal twist on Hobbes offered a solution to the dismaying political reality that succeeded the Restoration. After the fountains stopped foaming with beer and the maypoles came down, Charles II and his brother were revealed as perilously popish sovereigns for a Protestant state. Locke's cautious account of political, psychic, and moral freedom struck a new note in the philosophy of pain and pleasure, setting the terms of the British Enlightenment for the next hundred years.

THE GREATEST HAPPINESS FOR THE GREATEST NUMBER

In the century between the Glorious Revolution and the French Revolution, British philosophers started to argue for a species of ethical bookkeeping. In order to decide what to do, we should try and foresee the consequences of our actions and then tot up the pleasure and pain sides of the ledger. In the 1720s the influential Glasgow professor of moral philosophy Francis Hutcheson argued that actions are good or bad in proportion to the quantity of happiness or misery that flow from them: "*That action* is best, which procures the *greatest happiness* for the *greatest numbers*; and *that worst*, which, in *like manner*, occasions *misery*."[39] He even went so far as to reduce these utilitarian, cost-benefit calculations to actual equations: "goodness = benevolence × ability," and other such formulations. "Applying mathematical calculation to moral subjects will appear perhaps at first extravagant and wild," he suggested, little anticipating how extremely successful this strategy was to become when economists adopted it a hundred and fifty years later.[40]

Hutcheson's younger compatriot David Hume, librarian at the University of Edinburgh (among other meagerly compensated posts), followed suit. Our instinctive approval of "justice, allegiance, chastity and good manners" lay in the "utility" that flowed from them. For Hume, it was the overall pleasurable consequences of certain actions and tendencies that explained why we count them good. Utility, he claimed, was "the *sole* source of that high regard paid to justice, fidelity, honour, allegiance, and chastity . . . humanity, generosity, charity, affability, lenity, mercy, and moderation."[41]

It was not long before the utilitarian creed recruited its first Continental adherent, in the person of Claude Helvétius, whose career as a tax farmer allowed him to retire early and devote himself to the cause of becoming the Newton of ethics. Helvétius's *De l'Esprit* of 1758 restated Hobbesian egotism with ringing clarity, reducing every emotion, aspiration, and deed of man to the search for pleasure (mostly sex and money) and the avoidance of pain. These were, according to his scheme, "the simple principles, on the unfolding of which depends the order and happiness of the moral world."[42] The book was condemned as "atheistic, materialistic, sacrilegious, immoral and subversive by all the authorities," but the intervention of Madame de Pompadour prevented Louis XV from actually punishing its author.[43] The consequence of the book's scandalous reputation was, naturally enough, wild success: it was translated into every major European language and sold many thousands of copies.

In 1764 a young Italian admirer of Helvétius, Cesare Beccaria, published a passionate utilitarian argument for the abolition of capital punishment and torture. An ideal compact of free men, he declared, would devise laws with only one aim in view, that of *"the greatest happiness of the greatest number."* He proposed that a universal morality could be deduced from the fact that "pleasure and pain are the only springs of action in beings endowed with sensibility." It followed that "crimes are only to be measured by the injury done to society." Accordingly, punishments should be devised that "will make the strongest and most lasting impressions on the minds of others, with the least torment to the bodies of the criminal."[44] Hobbes had already argued that we should care about deterrence rather than retribution. In the same vein, Beccaria urged that pain to the criminal should be whittled down to the absolute minimum compatible with putting other people off criminal activity. But while Hobbes's proto-utilitarian ethos was in the service of returning England to absolute sovereign rule, Beccaria's was in the name of a more enlightened future.

MORAL SENTIMENTS

Meanwhile, in Edinburgh, David Hume had begun to collaborate with a budding philosopher, twelve years his junior, by the name of Adam Smith. A bachelor, whose widowed mother was "from first to last the heart of [his] life," Smith was the original absent-minded professor, known for talking to himself when alone and for his frequent lapses into complete abstraction in company.[45] One famous story has him pouring tea over some bread and butter and then drinking it, before declaring it the worst tea he had ever tasted. He could recite by memory vast screeds of poetry in many languages, and had an extraordinary command over a wide range of subjects, but to the end of his life he wrote a childish hand, and he always complained of difficulties in composition. The pains he took paid off handsomely. His books are still eminently readable, claimed as its own by every new generation of theorists of capitalism, on both the left and right. Karl Marx was among his admirers, while the Conservative British prime minister Margaret Thatcher was said to carry around his 1776 *Wealth of Nations* in her legendary handbag.

Smith was also the author of the lesser-known work *The Theory of Moral Sentiments*, published in 1759, which he regarded as his best and most important book. His whole system of moral psychology was based on a couple of simple observations about human nature. For Smith, we are, above all, social beings, who long for the affection and admiration of our fellows. We are also rational animals, with an innate capacity to mentally exchange places with other people. This ability to see the world through others' eyes functions as an internal regulator of our behavior. By allowing us to view ourselves as others see us, it prompts us to modulate our actions in accordance with the demands of social exchange. Although we are a world away from Hobbes's quarrelsome little man, Smith's Enlightenment individual has something of the same vainglorious and pleasure-seeking qualities. We are impelled to behave decently, Smith suggested, by our intense desire to be liked and respected, while we engage just as relentlessly in a quest for material self-betterment as a result of our concomitant need to be admired and envied.

Smith called our capacity to inhabit other perspectives the "impartial spectator" and the "demi-god within the breast." Although we are naturally susceptible to overweening self-love and arrogance, the impartial spectator's view of ourselves administers a much-needed dose of realism: "When he views himself in the light in which he is conscious that others will view him, he sees that to them he is but one of the multitude in no respect better than any other in it." The book can profitably be read as an Enlightenment version of *How to Win Friends and Influence People*. Smith argued that the more rigor, humility, and insight we bring to the task of attending to the impartial view of ourselves, the more we will learn to tune our behavior to the people around us, and the more virtuous and beloved we will become: "If he would act so as that the impartial spectator may enter into the principles of his conduct . . . he must, upon this, as upon all other occasions, humble the arrogance of his self-love, and bring it down to something which other men can go along with."[46] Surely this still counts as thoroughly good advice.

Smith's differences with Hume were subtle but far-reaching, pivoting around the conflict between utility and justice. To make the point, he narrated the melancholy fate of a sentinel who had fallen asleep on the watch and been put to death, a real-life example that must have made a great impression on him, as he returned to it at least three times in different contexts.[47] On the grounds of utility, the case was clear-cut: such a man "suffers death by the laws of war, because such carelessness might endanger the whole army." On the other hand, "the natural atrocity of the crime seems to be so little, and the punishment so great, that it is with great difficulty that our heart can reconcile itself to it."[48] Smith thought that we should recoil from the utilitarian calculus, even as we allow the tough decisions that are sometimes made in its name. With *The Theory of Moral Sentiments*, utilitarianism was beginning to butt up against its own emotional limitations.

TASTING THE FELICITY OF ANGELS

Fittingly, it was an English medical doctor who turned pain and pleasure into a full-blown neurological and physiological theory. In 1749 the physician and mathematician David Hartley laid out a materialist analysis of human psychology. He was a determinist, for whom thought was a physical thing, the product of minuscule vibrations in the nervous system. Describing how complex ideas and actions were built up of simple stimuli, Hartley describes how a child's grasp reflex evolved into a "Power of obtaining Pleasure and removing Pain."[49] Mostly forgotten today, his developmental account of pain, pleasure, and action was hugely influential in its time.

One of Hartley's followers was the British chemist and philosopher Joseph Priestley, inventor, among many other accomplishments, of laughing gas and the carbonated beverage. In 1768 Priestley published his *Essay on the First Principles of Government*, which articulated the relation between happiness and social organization in utilitarian terms: "the good and happiness" of the majority of the members of any state "is the great standard by which everything relating to that state must finally be determined." But for Priestley, there was no contradiction between this goal and man's "natural right of relieving himself of all oppression, that is, from every thing that has been imposed upon him without his consent."[50] A rights-based political system would allow the sacred individual—whom he described as a man swaddled in domestic bliss with a devoted wife and children—opportunities for unlimited happiness. "We are perpetually deriving happiness from new sources," Priestley declared, and even before we depart this world we "are capable of tasting the felicity of angels."[51]

Theologian, historian of science, political theorist, and revolutionary chemist, Priestley articulated a philosophy of happiness that was universalistic, utopian, and bursting with faith in scientific and spiritual progress. Just as the maturing individual attained ever-higher reaches of happiness, so the maturation of science and government would one day do the same for the whole species: "The end will be glorious and paradisiacal, beyond what our imaginations can now conceive."[52]

Jeremy Bentham devoured Priestley's *Essay* right after it was first published, but he gave the radical chemist's utilitarianism a neo-Hobbesian twist. Far from proposing that rights ensured happiness, Bentham asserted that they were replete with dangers. For him, as for Hobbes, the assertion of rights and liberties could lead to anarchy and violence. But while Hobbes had accepted the existence of natural rights and argued that self-preservation must trump them, Bentham went a step further and denied their existence altogether.

GENIUS IN PETTICOATS

As befits a theorist of happiness, Jeremy Bentham was the progeny of pleasure. His father Jeremiah disappointed his snobbish parents by falling in love with the daughter of a prosperous tradesman, whom he met at Buckholt Acres, a place of entertainment near Epping Forest. The marriage was a great success, and Jeremiah averred that he lived "in a constant and uninterrupted state of nuptial happiness." The merry couple set up house in Red Lion Street, Houndsditch, in the East End of London where, four years after the wedding, the first son, christened Jeremy, was born on February 15, 1748.

Bentham was a tiny child, "the feeblest of all feeble boys." He liked to tell the story of the ill-fated dance lesson inflicted on him at the age of seven, in which he proved so weak as to be unable to support himself on tiptoe. The teacher mortified him by asking "whither my calves had gone a'grazing." He was as emotionally sensitive as he was physically frail: his earliest memory was of bursting into tears when he perceived his mother to be saddened by his refusal of a sweet. He characterized this as the "pain of sympathy," Adam Smith's language for the way in which empathy awakens the moral sense in man. From infancy he abhorred all games and sports involving animal suffering.[53]

Mr. and Mrs. Jeremiah Bentham proudly documented the freakish intellectual prowess of their tiny child. No biographical sketch of the man fails to recount the occasion when he was out on a walk with his mother and father, ran home ahead of them, and was found upon their return reading a vast *History of England* at the dining room table. This event occurred before he was three and a quarter years old, remarked because he still had not been "breeched," that rite of passage when a little boy was presented with his first pair of trousers. "The tale was often told in my presence of the boy in petticoats, who had come in and rung the bell, and given orders to the footman to mount the desk upon the table, place the folio upon the desk, and provide candles without delay."

The pleasure Bentham's father took in his precocity is also evident in the line of Latin preserved in a scrapbook with the caption: "The line pasted herein was written by my son, Jeremy Bentham, the 4th of December 1753, at the age of five years, nine months, and nineteen days." Such was the rapidity of the little boy's intellectual development, his proud parents had to measure it in days. His first lessons were in Latin, Greek, and music, and at the age of seven he was provided with a French tutor, developing such facility with the language that he later claimed to write it more fluently than English.[54]

CRITIC OF WHIG CONSTITUTIONAL LAW

Young Jeremy burned quickly through his formal schooling, attending a series of venerable institutions, each exemplifying the eighteenth-century degradation of English education. At Westminster School, he avowed, his most valuable lessons were garnered from the little boy who shared his bed, who told him stories peopled with heroes and heroines whose virtues he resolved to emulate. A year after the death of his mother, at the age of twelve, he matriculated at Queen's College, Oxford, where he judged the tutors "profligate," "morose" and "insipid." At fifteen he obtained his bachelor's degree and began his formal legal training by the expedient of taking meals at the Lincoln's Inn, an ancient society of lawyers, whose system of education had fallen apart with the English Civil War, to be replaced by perfunctory dining rituals and the recitation of empty legal formulae. The following year he went back to Oxford to attend the lectures of William Blackstone, the leading lawyer in England, whose successful private course was perhaps the best that legal education had to offer. As he listened, Bentham burned with adolescent contempt for the shoddiness and complacency of Blackstone's arguments for the moral basis of common law.[55]

In 1766 Bentham received his master's degree and strutted about in his new gown "like a crow in the gutter." In the end it was a coffeehouse library that delivered to him the intellectual inspiration so lacking in Oxford's official curriculum. Returning to that cloistered city after receiving his MA, he happened to visit an old haunt, the Harper's Coffee House near Queen's College. Through the circulating library attached to this institution, Bentham laid his hands upon Joseph Priestley's 1768 *Essay on the First Principles of Government*. He later described the experience of reading the essay as a revelation, and attributed to Priestley his discovery of the greatest happiness principle as the overriding natural law of political philosophy.[56]

Like many people destined to make their mark upon history, he rebelled against the course set out for him by his father. He hated the law, and burned only with a zeal to reform it. The mountain of seventeenth-century legal "trash" he was expected to know filled him with nothing but despair, and he could never bring himself to argue a case. In October 1772 he wrote to his father declaring his inability to practice his chosen profession and his intent to devote his time to critiquing it. His father had settled some property on him a few years earlier, and thus conferred the independence required for this course of action. Bentham continued to live at Lincoln's Inn and pursued a life of great simplicity, rising early and going for a long walk, followed by a solitary breakfast, after which he would work until four in the afternoon. Every evening he dined with his father.[57]

Writing came hard to Bentham in those days. He began sentences which he could not complete, wrote down words and phrases on scraps of blotting paper, which he kept in drawers in the hope that he might be able to form them into continuous prose in happier times. He likened committing a phrase to paper to screwing a piece of wood in a vice, to be planed and polished at leisure. The results of his efforts do not make for easy reading, but he ended up changing the world by changing the English language, as evidenced by the alphabetical list of his neologisms that has been placed online by helpful historians at University College London. Those that caught on constitute an index to modernity, the rest a charming satire on eighteenth-century intellectual eccentricity. In the former category are maximize, minimize, rationale, demoralize, deontology, disambiguation, dynamic, international, unilateral, exhaustive, self-regarding, cross-examination, sexual desire, and false consciousness. Neologisms that failed to stick include metamorphotic, disceptatorial, morthoscopic, undisfulfilled, infirmation, subintellect, thelematic, antembletic, scribblatory, imperation, and incognoscibility.[58] Based on Bentham's record to date, some of these may yet enjoy their day in the sun. Scribblatory, anyone?

In 1776, at the age of twenty-eight, Bentham finally went to press with his first considerable work, a criticism of the *Commentaries* of William Blackstone, the lawyer whose Oxford lectures he had so despised as a teenager. Drawing upon his utilitarian predecessors, including Helvétius, Beccaria, and Priestley, Bentham asserted that the fundamental axiom of moral science had been established: "It is the greatest happiness of the greatest number that is the measure of right and wrong."[59] Bentham explained that pain and pleasure were the sole ground for the natural laws of ethics and jurisprudence. These alone will command universal recognition and "firmly fix the attention of an observer." "Utility" was defined as that tendency in any social arrangement that conduces to happiness; "mischievousness" was Bentham's term for the opposite.[60]

The measure of utility then became Bentham's platform from which to demolish Blackstone's Whig apologia for English law. Especially contemptible to the young philosopher was Blackstone's invocation of the Glorious Revolution. According to Blackstone, William and Mary's Act Declaring the Rights and Liberties of the Subject was an original compact, made between monarch and people in 1688, in which obedience to the sovereign was offered in exchange for security and happiness. This

"Whig-Lawyer fiction" was the object of relentless derision on Bentham's part, who dismissed it as nothing more than a chimera designed to support an indefensible miscellany of laws and customs that operated to the sole advantage of lawyers and judges. As for Blackstone's concept of rights, Bentham derided it as meaningless and then spent a few pages demonstrating how much more straightforward everything became once utility had been substituted for it: "Why turn aside into a wilderness of sophistry, when the path of plain reason is straight before us?"[61]

Bentham was caught unawares by how subversive and inflammatory his critique seemed, especially to members of his own Tory party, whom he expected to join him in ridiculing Whig pretensions. As he told his executor in later life, "I . . . never suspected that the people in power were against reform. I supposed they only wanted to know what was good in order to embrace it."[62] In 1789 he responded to being called "dangerous" by the solicitor general with a charge of self-refutation: "When a man attempts to combat the principle of utility, it is with reasons drawn, without his being aware of it, from that very principle itself." Eventually he came to realize that it was the *universality* of the principle of utility that the powerful found offensive, and that his critics were actually perfectly consistent in pursuit of their own advantage. In 1822 he admitted that the solicitor general's comment was far from self-contradictory and that the principle of utility was, indeed, "*Dangerous* . . . to the interest—the sinister interest of all those functionaries, himself included, whose interest it was to maximize delay, vexation, and expense, in judicial and other modes of procedure, for the sake of the profit extractible out of the expense."[63]

In 1776, on the eve of the American Revolution, Bentham was that most contradictory of creatures, a Tory radical. After the American colonists declared their independence from Britain, Bentham produced a "Short Review on the Declaration" dismissing the founding document of the new American republic as "contemptible and extravagant." His first line of attack was to point out the contradictions in the colonists' actions: "If the right of enjoying life be unalienable, whence came their invasion of his Majesty's province of Canada?" From this he moved to a dismissal of the whole notion of unalienable rights, on the grounds that they render impossible the maintenance of society. If "all penal laws . . . which affect life or liberty are contrary to the law of God, and the unalienable rights of mankind," then "thieves are not to be restrained from theft, murderers from murder, rebels from rebellion."[64] In the *Fragment of Government*, the subversive and revolutionary quality of Bentham's greatest happiness principle was everywhere apparent. Faced with the American colonists' break from the Crown, the principle of utility turned into a powerfully comprehensive, physically concrete alternative to the rhetoric of rights.

SOVEREIGN MASTERS

A few short years after the American Revolution, the Whig aristocracy gently gathered Bentham to its soft white bosom. In 1781 William Fitz-maurice, Earl of Shelburne, made a visit to Bentham's garret at Lincoln's Inn. An invitation to Bowood, Shelburne's beautiful stately home in Wilt-shire, followed. There, Bentham spent the happiest weeks of his life. He was, in his own words, "caressed and delighted" by the grace and elegance of Whig aristocratic life. "People here," he wrote incredulously to a friend, "do just what they please."[65] He fell in love with Caroline Fox, Shelburne's fourteen-year-old niece, and many years later asked her to marry him, an offer she refused. He rubbed shoulders with the great and the good, playing chess with William Pitt, the future prime minister, and Lord Chatham, his brother. He had brought printed proofs of a manuscript over which he had been laboring, and in the evenings Lord Shelburne read portions of the text to the ladies.

Emboldened by the sweetness of his reception at Bowood, Bentham spent the next decade looking for a grand outlet for his reformist zeal. In 1785 he set off to join his much younger brother in Russia. (Of seven children, Jeremy was the oldest and Samuel the youngest. They were the only two who survived into adulthood.) Colonel Bentham, as Samuel styled himself, was living on the estate of Prince Potemkin, a favorite and one-time lover of Catherine the Great, and one of the most powerful men in Russia. As governor general of the New Russian colonies in the south, Po-temkin reigned like a second emperor over a vast territory annexed from the Ottomans. Samuel Bentham was appointed as an engineer and put in charge of the prince's sprawling estates in Belorussia. Jeremy arrived in 1786 and stayed for a year. He had hoped to interest the empress in his schemes, but the contact was never made. He took one long look at what he described as a "Bedlam" of debauched British expatriates, promptly moved into an isolated cottage, and spent his days much as he had in London, rising early and laboring all day on his manuscripts. The fruits of his Russian sojourn were his design for the Panopticon Penitentiary and his first pamphlet to meet with critical success, *Defense of Usury*, arguing for the utility of financial capitalism (which Adam Smith had attacked in *The Wealth of Nations*) and against Britain's anti-Semitic laws prohibiting the lending of money at interest.

In 1787 Bentham returned to England and moved into a little house on a farm in Kent, playing the harpsichord, eating the farmer's wife's excellent cooking, "seeing nobody, reading nothing, and writing books which nobody reads," in the verdict of one friend.[66] The proofs that Shelburne had read aloud to the assembled company at Bowood languished unpublished for nearly a decade. Bentham had tried to rewrite a portion of the manuscript and found himself "entangled in an unsuspected corner of the metaphysical maze." Predictably enough, "suspension brought on coolness, and coolness . . . ripened into disgust." He was finally goaded by the circulation of pirated versions into publishing it. He added a postscript and brought the book out in 1789 as *An Introduction to the Principles of Morals and Legislation*. It opens with the famous formulation of the psychology that underpins utilitarian ethics: "Nature has placed mankind under the governance of two sovereign masters, *pain* and *pleasure*."[67]

"The business of government," Bentham baldly declared, "is to promote the happiness of the society, by punishing and rewarding."[68] Beccaria had advanced the same principle for the reform of vengeful and disproportionate laws, but the Milanese philosopher did not have Bentham's stamina for worrying away at the details. By erecting a standard against which the legitimacy of laws could be judged and then proceeding, inexhaustibly, to judge them, Bentham was indeed, in Thomas Macaulay's words, "the man who found jurisprudence a gibberish and left it a science."[69]

As part of his systematizing efforts, Bentham set out to crown utilitarianism as the sole sovereign morality, eliminating all other pretenders to the throne of ethical significance. This Herculean act of philosophical hygiene came close to nihilism. Nothing was to be judged good or bad in itself, least of all human motivations and dispositions. Since pleasure was the only thing good in itself, and pain the only thing inherently evil, any given human action was to be judged solely on its consequences; motivations and intentions were, in themselves, morally neutral. Our vocabulary had therefore to be leeched of its dusty evaluative overtones. For "lust" Bentham substituted "sexual desire"; in place of "avarice" he suggested "pecuniary interest":

> 1. A man ravishes a virgin. . . . 2. The same man at another time exercises the rights of marriage with his wife. In both cases, however, the motive . . . may be neither more nor less than sexual desire. . . . 1. For money you gratify a man's hatred, by putting his adversary to death. 2. For money, you plough his field for him.—In the first case your motive is termed lucre, and is accounted corrupt and abominable: and in the second . . . it is styled industry and is looked upon as innocent at least, if not meritorious. Yet the motive is in both cases precisely the same: it is neither more nor less than pecuniary interest.[70]

Bentham confessed himself so indebted to his liberal predecessors that he seems not to have recognized how radical his utilitarianism had become. He had swept away Hume's respect for traditional virtues, Beccaria's Enlightenment sentimentalism, and Priestley's natural rights. In pairing murder with farming and rape with marital sex, Bentham exposed the vein of moral nihilism in utilitarianism that would agitate its critics from his day to ours.

In place of tradition, sentiment, and rights, Bentham insisted on the brute physicality of pain and pleasure. To that end, he devoted a whole chapter to the enumeration of "Circumstances Influencing Sensibility." By suggesting that "Health," "Strength," and "Hardiness," as well as the "absence of disease," the "firmness of the muscular fibers," and the "callosity of the skin," were important for ethics, Bentham brought politics onto the same terrain as medicine. The connection was literal: the measure of a good or bad law was how much pain or pleasure it caused, and these effects were mediated by medical facts about physical vigor. In a footnote Bentham confessed that the subject was "one of the most difficult tasks, within the compass of moral physiology."[71] What he seemed not to have noticed, in his handwringing about the complexity of the undertaking, was that "moral physiology" was a whole new discipline.

By this time, a golden opportunity to enact sweeping reforms had opened up across the Channel. Already in 1788 Bentham foresaw that a new world was being hatched, in need of a fresh legislative procedure. He dashed off a series of pamphlets, making suggestions for the constitution and practice of the French National Assembly. These were received by the revolutionaries with varying degrees of warmth. An enthusiastic response to his plan for a new judicial system did not result in its enactment. Instead, the Revolution seemed hell-bent on hurtling leftward, to Bentham's dismay. The launching of the Revolutionary Wars, the legislative assaults on private property, the September Massacres, and the Terror appalled him, as they did most of the Revolution's sympathizers in his circle.

In the autumn of 1792 Bentham was still describing himself as "a royalist in London" and "a republican in Paris,"[72] but the regicide of Louis XVI in January of the next year saw him inveighing against the Revolution and attacking once again the pernicious doctrine of natural rights. In a series of works culminating in a famous rant against the French Declaration of Rights he worked out the differences between the measure of utility and the doctrine of natural rights—between bloodless, rational, utilitarian reform and violent, irrational, anarchical insurrection: "Natural rights is simple nonsense: natural and imprescriptible rights, rhetorical nonsense—nonsense upon stilts."[73]

HOUSES OF INDUSTRY

Unable to save the French from their folly, Bentham threw his energies into building his most concrete alternative to revolution, the Panopticon Penitentiary. He had returned from Russia in 1787 with the Panopticon plans in his luggage. In his design, the prison warder was housed in a central chamber, around which the inmates' cells were arrayed in a large circle. Windows behind each cell silhouetted the prisoners during the day, while lamps illuminated them at night. During the hours of daylight, the interior of the central chamber was naturally invisible, like the dark rooms of a house viewed from the street. A system of blinds and grills kept it that way at night. This unidirectional flow of photons was intended to stop the prisoners from knowing when they were being watched. It would act as an invisible guard: a vigilant eye regarding the prisoner's silhouette, ceaselessly checking on his behavior, inducing him day and night to discipline *himself*.[74]

With characteristic exhaustiveness, Bentham wove a tapestry of further details around his central conceit, prescribing the shape of the iron trays on which the prisoners' potato diet would be served, the solemn organ music that would accompany their first bath, the fabric of the guards' uniforms, the layout of the gardens, the mechanism of the ventilation system, the design of the latrines. At this early stage, Bentham's Panopticon Penitentiary was a fantasia of glass and iron, at the center of which the manager's chamber crouched like a dark spider in a web of light. It was also a counter-Enlightenment variation on Adam Smith's notion that our social behavior is guided, above all, by our inner awareness of how we appear to others. The Panopticon took this idea of a controlling gaze and externalized it. Turning Smith's "demigod within" into the design for a physical building, Bentham exchanged the inner light of human reason for an outer light directed at the body's transgressions.

When the French Revolution broke out, Bentham tried to interest the new French National Assembly in his prison design. The politicians who initially expressed their enthusiasm subsequently proved elusive. Bentham then turned to the Scottish Parliament and the Irish government. They strung him along with half promises and long delays. In January 1791 he wrote to William Pitt, the British prime minister, offering his services as the Panopticon's builder and manager. When he did not hear back from Pitt immediately, his obsession grew. He snowed his wide acquaintance with copies of his proposal. He suggested pedantic amendments, detailed budgets, and inspiring slogans ("Mercy, Justice, Vigilance").[75] He wheedled and begged and joked. His tone became bitter—"You grew tired of it," he wrote to the Irish chancellor of the exchequer. "I did the same. Ennui is catching."[76] In November 1791 he wrote to Pitt again, enclosing another copy of the proposal. Still nothing. Two months later, he wrote yet again: "I am now ready to execute the plan stated . . . *taking on myself all expence of building*."[77]

In March 1792 Bentham's father died and he inherited a comfortable living, as well as the family house in Queen's Square Place in Westminster. Fired with fresh optimism, he pressed on with the Panopticon campaign and was rewarded with a visit from the secretary of state, who came to the house and "admired or pretended to admire" a model of his design.[78] In 1794 Parliament actually passed a bill requisitioning a site in Battersea Rise, southwest of London. To no avail. The owner, the second Earl Spencer, kept up a campaign of steady stonewalling that eventually saw Bentham off.[79] Another site was similarly thwarted when its owners objected that the Panopticon would overlook the path of the family matriarch's favorite daily stroll. The pious and parsimonious Viscount Belgrave blocked a proposal to build it on wasteland in Pimlico. Eventually Bentham gave up. In 1797 he noted bitterly that "the claims of justice and utility" were helpless in the face of "favour and connection." From "long and dear bought experience," he had discovered that nothing "can stand against sinister influence."[80]

Bentham's reaction to the Panopticon debacle was symptomatic of his messianic streak. He took the thwarting of his pet scheme by a series of skeptical aristocrats to be evidence of widespread human corruption. Smarting with failure, he now bitterly acknowledged the "sinister influence" lurking in every nook and cranny of human institutions. The law was an incomprehensible mess because of the self-interest of lawyers, who profited from ordinary people's inability to navigate the labyrinth. The same selfish motives were rampant among members of Parliament and the aristocracy. With his descent into cynicism, Bentham's social engineering became much more psychologically elaborate. Basing his schemes on the premise of primal egotism, he devised ever more complex social machinery for transforming human self-interest into social utility.

For the construction of prisons, asylums, and workhouses, Bentham decided that the key to lasting success lay in what he liked to call "pecuniary interest." In 1798 he published a grand scheme for harnessing the power of greed to address the problem of poverty. His *Outline of a Work Entitled Pauper Management Improvement* proposed that five hundred Panopticon "Houses of Industry" be built in a regular grid across the nation, to accommodate a total of half a million souls.[81] To arouse the self-interest of the managers of the system, the Houses of Industry would be run by a for-profit "National Charity Company," loosely modeled on the East India Company, which would raise money from shareholders and then disburse profits from pauper labor.[82] In addition to the opportunity to exploit a free labor source, Panopticon managers would earn money from a life assurance scheme, receiving a fixed payment for every year of life of every child in their care, and being fined for the death of every woman in childbirth. Self-interest, Bentham declared, was the only reliable basis for any scheme: "a system of economy built on any other foundation, is built on quicksand."[83]

A far humbler version of self-interest was marshaled to make the paupers comply. The incentives for the inmates consisted of foodstuffs beyond the basic bread and water, less onerous labor, some "distinction in dress." These humble ameliorations would be exquisitely manipulated to squeeze every last drop of utility from their bodies: "Not the motion of a finger—not a step—not a wink—not a whisper—but might be turned to account in the way of profit in a system of such a magnitude." Steam power would be rendered unnecessary in some industries, Bentham mused, as the work could be fueled instead by pauper sweat. Importantly, the diet of the workhouse should be controlled so that "Charity maintenance—maintenance at the expense of others, should not be made more desirable than *self*-maintenance." This last idea—that life in the workhouse should be markedly more unpleasant than life outside— became one of the planks of the utilitarian reforms to the Poor Laws, discussed in the next chapter.[84]

Bentham's vision of pauper management was a vast manufactory of social reform in which pains and pleasures were to be managed in every detail of architecture, organization, and administration. According to the strictest accounting of utility, paupers, criminals, and the insane were to be saved, but in exchange for bare life they would work. Rationally governed through their humble appetites and aversions, they would be goaded and cajoled into laboring in the name of the greater good. From the monetary penalties and payments held out to the contract managers, to the pathetic rewards and tedious punishments dealt out to the inmates, the whole system, in all its exhaustive ramifications, was predicated on the idea that humans could be completely manipulated through pain and pleasure. Bentham chained Epicurus to a walking wheel, to circle ceaselessly at the center of his machinery for turning hunger into happiness.

The Houses of Industry scheme was the apogee of Bentham's illiberal and controlling tendencies. It also marked a turning point in his career. Thwarted in his plans by aristocratic nimbyism, he subsequently redirected his reformist zeal against the class that had frustrated his ambitions. In the case of parliamentary politics, he came to believe that the solution lay in secret ballots and universal suffrage, making the power of the rulers directly dependent upon the favor of the ruled.[85] This Janus-faced political attitude—universal suffrage on one side, behavior control on the other—was the legacy that Bentham bequeathed to the reformers who acted in his name. At the very top of the social pyramid, parliamentary reform would expose the corrupt actions of the rich and the powerful, aligning the actions of government with the interests of the majority. In the middle stratum—exemplified by Panopticon managers—inviolable property rights and universal pecuniary interest would secure the comforts of bourgeois citizens. For the hapless paupers at the bottom of the hierarchy, the prods and goads of utility came in the form of food rewards and work relief. Thrust beyond the pale of liberal freedoms by the harshness of their predicament, the poor were to be manipulated in the spirit of training nonhuman animals. The Panopticon was never built, but its techniques of pauper management left their imprint on British scientific psychology. In the end, Bentham's design was more faithfully and fruitfully realized in the behavior laboratory than in any prison or asylum.

The next chapter pairs Bentham with his fellow utilitarian Thomas Malthus. Even more than Bentham, Malthus urged that the pauper on the brink of starvation represented the animal state of the human species. Hunger, for Malthus, defined the limit of civilization, revealing that man was engaged in a brutish struggle for existence, a Hobbesian war of all against all, in which the strong would survive and the weak would perish. With this move, he supplied nineteenth-century science with a deep biological level to utilitarian analysis. Between them, Malthus and Bentham constructed the utilitarian self in the image of the workhouse inmate. Its body was Malthusian: a bare, forked creature, hovering on the brink between procreative life and premature death, threatening to breed, competing for scarce resources. Its nervous system was Benthamite: an animal bundle of appetites and aversions completely controllable by reward and punishment.

3

Nasty, British, and Short

THE LESSON OF HUNGER

While insurrection blazed in France, industrialization reached escape velocity in Britain. Aristocrats enclosed the last tracts of the medieval commons, replacing subsistence farming with private parklands and large-scale agriculture. Traditional crafts were taken over by machine production and the factory system. A great mass of men, women, and children were set on the move, migrating in their thousands to work in urban factories, to become itinerant farm laborers, or to extract coal from the new deep-pit mines.

To sensitive observers, it seemed as though the very sources of Britain's climbing prosperity—coal power, factory production, a free market in human labor—consigned the poor to the harshness of animal existence. The blackened faces, burning eyes, and sickly offspring of industrial workers presented a living reproach to flamboyant displays of bourgeois luxury.[1] Hundreds of anguished dissertations about inequality were penned toward the end of the eighteenth century. "Our nobility live in palaces, our gentry live in villas, commerce has made us a nation of gentry," commented one eloquent witness in 1775, before lamenting the terrible sight of "our starving, naked, unshelter'd, miserable Poor."[2] In the malnourished body of the British pauper, the emerging laws of economics seemed to confront immutable laws of organic existence. How were the new middle classes to obey the imperatives of self-enrichment without causing mass mortality by overwork, occupational illnesses, and starvation?

In his 1776 *Wealth of Nations*, Adam Smith proposed that hunger was simply the result of too many mouths to feed: "Every species of animals naturally multiplies in proportion to the means of subsistence, and no species can multiply beyond it. But in civilized society it is only among the inferior ranks of people that the scantiness of subsistence can set limits to the further multiplication of the human species." The bottom of the socioeconomic scale, in other words, represented a brutal margin between civilization and nature, defined by starvation. Having presented the reader with this melancholy thought, Smith then dismissed it with a cheerful wave, proposing that the market would take care of the problem: "Demand for men, like that for any other commodity, necessarily regulates the production of men; quickens it when it goes on too slowly and stops it when it advances too fast."[3]

The Wealth of Nations was composed during a heady period, after the Seven Years War, when Britain had wrested control of vast tracts of North America from the French and begun to see the whole globe as potentially within its imperial grasp. Unfortunately for the march of British power, the publication of Smith's famous book coincided with the beginning of the American Revolution. In the gloomy aftermath of the British defeat, some of Smith's successors became less sanguine about the market's capacity to defeat hunger, crime, and insurrection. Maybe, they surmised, the facts of procreation and starvation obeyed older, deeper rhythms. Perhaps the human breeding cycle was slower than the fluctuations of supply and demand, putting "the production of men" out of step with market conditions, and leaving the job of population control to the more elemental forces of famine and disease.

The first to explore the darker side of surplus population was a friend of Bentham's. In 1781 Bentham used the word "utilitarian" as a noun for the first time, in a letter from Bowood House, applying it to a fellow guest, the Reverend Joseph Townsend: "a very worthy creature . . . his studies have lain a good deal in the same track with mine. He is a utilitarian, a naturalist, a chemist, a physician."[4] Famous for his proprietary remedy for syphilis, Townsend was a dedicated seeker after the natural laws of human happiness. As Bentham remarked in another letter a few days later: "What I have seen of him, I like much; his thoughts have run pretty much in the channels mine have run in."[5]

In 1786 Townsend published an analysis of the problem of poverty, in which he argued against poor relief on the grounds of the disciplining effects of hunger: "Hunger will tame the fiercest animals, it will teach decency and civility, obedience and subjection, to the most brutish, the most obstinate and the most perverse."[6] He recounted the story of a deserted island on which a Spanish explorer placed a pair of goats. The goats bred until their descendants filled the whole island. Up until this critical moment, Townsend declared, the creatures were "strangers to misery and want and seemed to glory in their numbers." But as soon as they had reached the population threshold, "they began to suffer hunger." Now the balance had to be violently regained: "In this situation the weakest first gave way, and plenty was once again restored."

Warming to his theme of population overload, subsistence, competition, and extinction, Townsend continued his parable. After realizing that English sailors were feeding off his goats, the Spaniard put a greyhound dog and bitch on the island, in the hope that they would eat the goats, have lots of goat-eating puppies, and deprive his English enemies of their dinner for all time. The goats promptly retired to the high elevations where the dogs could not follow them. Now only "the most watchful, strong, and active of the dogs could get a sufficiency of food. . . . The weakest of both species were among the first to pay the debt of nature; the most active and vigorous preserved their lives." This preliminary sketch of a mechanism regulating population, competition, and food supply was eventually formulated as the law of natural selection. And just like the social Darwinists of a later age, Townsend took the moral of his fable to be about human society: "It is the quantity of food which regulates the numbers of the human species."[7]

At the same time as Townsend's foray into political economy, the Anglican archdeacon of Carlisle, William Paley, published a defense of utilitarianism that exactly caught the national mood about poverty and property. Paley's *Principles of Moral and Political Philosophy* contained a clear, concise, and amusing summary of utilitarian principles, deduced without any philosophical hairsplitting, from the benevolence of the deity. A section on property rights opened with a startling parable concerning a flock of pigeons surrounded by seeds. Paley exhorted the reader to imagine that ninety-nine of the birds, instead of just pecking at the food wherever and whenever they liked, worked assiduously to pile the seeds into a big heap, "reserving nothing for themselves." The heap was kept "for one, and that the weakest and perhaps worst of the flock," who spent the whole winter devouring and wasting it, while the others made sure to tear to pieces any one of the ninety-nine who dared to touch a single grain of it. This bizarre behavior, Paley declared, was "nothing more, than what is practiced and established among men."

After setting the reader up for a radical denunciation of private property, Paley then proposed that "there must be some very great advantages to account for an institution, which in one view of it, is so paradoxical and unnatural." Conducting a quick survey of those advantages—"none would be found willing to cultivate the ground, if others were to be admitted to an equal share of the produce"—he concluded that "the balance . . . must preponderate in favour of property with manifest and great excess."[8] By defending social inequality with such verve and wit, he ensured the success of the book: adopted as a pillar of the curriculum at Cambridge University, and going into dozens of editions, it was considered a touchstone for British moral thought up to the 1830s.

OLD POP

It was one of Paley's followers who took the national anxiety about poverty, reframed it as a biological problem, and stamped his name on it for all time. Thomas Robert Malthus was a thirty-two-year-old Paleyite utilitarian, country curate, and Fellow of Jesus College Cambridge when he published his *Essay on the Principle of Population* in 1798. Hoping to crush forever the crazy optimism of the French revolutionaries, Malthus analyzed poverty at the most basic, biological level, forecasting a future of ineradicable hunger and misery at society's margins.

Here are Malthus's "two postulata": "First, that food is necessary to the existence of man. Secondly, that the passion between the sexes is necessary and will remain nearly in its present state."[9] From this Hobbesian pair of propositions—we need to eat and we like to have sex—Malthus deduced the continued existence of hunger, poverty, war, famine, and pestilence:

> The power of population is so superior to the power in the earth to produce subsistence for man, that premature death must in some shape or other visit the human race. The vices of mankind are active and able ministers of depopulation. They are the precursors in the great army of destruction, and often finish the dreadful work themselves. But should they fail in this war of extermination, sickly seasons, epidemics, pestilence, and plague, advance in terrific array, and sweep off their thousands and ten thousands. Should success be still incomplete, gigantic inevitable famine stalks in the rear, and with one mighty blow, levels the population with the food of the world.[10]

Against Paley, who thought that acts of charity were good for the soul, Malthus argued that poverty was only exacerbated by the paternalistic strategy of throwing money at the problem. The inexorable laws of supply and demand were the source of the difficulty. By enabling paupers to buy more food, poor relief drove up the price of basic nutriments. By encouraging them to procreate, it drove down the price of labor. With every payment, the gap actually widened for the poor between the cost of living and wages. The result? More hunger, disease, war, and vice. Malthus was a devout Christian, but as a utilitarian he appealed to the most elemental facts of human life—sex, childbirth, and starvation—to argue against Christian philanthropy. Results, outcomes, consequences—these were all that mattered. Generous impulses must be suppressed for the sake of the poor themselves. No moral or religious scruples must be allowed to stand in the way of a clear-eyed analysis. And no sentimental failure of nerve should get in the way of enacting a hard-nosed utilitarian solution.

The *Essay* was a polemic, written in haste in a remote curate's cottage, without access to a library of relevant works. After its publication, Malthus resolved to make a more empirical study of the population problem. In 1799 he went on a tour of Scandinavia, taking copious notes on the health and customs of the famously long-lived people of Norway and their hungrier, sicker Swedish cousins. In Norway, he found that parish elders in the agricultural villages forbade young men to take wives until they were economically self-sufficient. Perhaps as a result of this exposure to Norwegian family planning, he began to entertain the possibility of deliberate intervention in the biological predicament of human existence. Given the inexorable laws of our animal nature, he asked, might we not deploy our God-given reason to mitigate their effects, and turn pauper misery into bourgeois enjoyment?

In the second edition of his *Essay*, published in 1803, Malthus outlined his Scandinavian prescription for prosperity. His solution was to persuade the working class to postpone marriage and procreation. This would be achieved by simply explaining to them why "moral restraint" was in their best interests: "that the withholding of the supplies of labour is the only possible way of really raising its price, and that they themselves, being the possessors of this commodity, have alone the power to do this."[11] Postponement of marriage, he observed, had long been practiced by members of the middle class, who could see how their individual comfort and status would be drained by family life under a certain income level. The trick was to get the poor to understand that the stakes were the same at the bottom of the social ladder, albeit in a more collective, attenuated form. If they could be persuaded en masse to limit the supply of workers through voluntary celibacy, then they could drive up the price of labor to the height of a living wage.

Birth control and abortion were abominations to Malthus under the terms of his Christian faith, and so strict sexual abstinence was the only permissible recourse in the matter of family size. All of his other recommendations were utilitarian through and through. With every new edition of the *Essay*, he appealed more insistently to the "test" and the "touchstone" of consequences. Even the obligation to practice "moral restraint" (sexual abstinence) rested, according to Malthus, "on exactly the same foundation as our obligation to practice any of the other virtues, the foundation of utility."[12] Weighing consequences was, for Malthus, the only rational approach in the face of "necessity, that imperious all-pervading law of nature."[13]

Malthus's dismissal of the notion of natural rights was as emphatic as Bentham's. In the context of an attack on Thomas Paine's *Rights of Man*, he explained why the idea of a "right to subsistence" on the part of the unemployed was doomed to failure:

At nature's mighty feast there is no vacant cover for him. She tells him
to be gone, and will quickly execute her own orders, if he do not work
upon the compassion of some of her guests. If these guests get up and
make room for him, other intruders immediately appear demanding the
same favor. The report of a provision for all that come, fills the hall with
numerous claimants. The order and harmony of the feast is disturbed,
the plenty that before reigned is changed into scarcity; and the happiness
of the guests is destroyed by the spectacle of misery and dependence in
every part of the hall, and by the clamorous importunity of those, who are
justly enraged at not finding the provision which they had been taught to
expect.[14]

As a specimen of pleasure-pain psychology, "the happiness of the guests
is destroyed by the spectacle of misery and dependence" is as uncom-
fortably insightful as anything in Hobbes. The passage hit a raw nerve,
and Malthus was condemned from the right and the left alike, just as
his seventeenth-century predecessor had been. Although the paragraph
disappeared from all subsequent editions of the *Essay*, it provoked
an enduring portrait of Malthusian heartlessness, indelibly drawn by
Coleridge, Dickens, and Marx, to name but a few.

In 1804 Malthus married and had to give up his Cambridge fellowship. The following year he took up the country's first professorship in
political economy, at the newly opened East India College in Haileybury,
where the students were trained in Sanskrit, Persian, Hindi, law, and
history, in preparation for careers in imperial administration. Malthus
worked at the college and lived in Haileybury for the rest of his life,
known affectionately by his students as "Old Pop," a diminutive of "Population Malthus."

Malthus argued against *everybody*. To him, charitable Christians and
wild-eyed Jacobins were equally deluded in their sentimental insistence
that the poor must be compensated for their plight. He was accordingly
derided and denounced from every wavelength of the political spectrum,
but in the end he won the debate. The target of his reformist zeal was
the system of relief payments to the poor, which were tied to the cost
of bread and the size of families. This system had been fought for by
an alliance of Evangelical Christians and paternalistic Tories, who had
succeeded by the mid-1790s in establishing a living wage for workers in

times of unemployment. At first, Malthus's counterargument gained little official traction. After the defeat of Napoleon in 1815, however, poor harvests, high grain prices, famine, and unemployment sent the burden on the public purse soaring upward, and Malthus's principle of population proved suddenly irresistible. Malthus provided both an explanation for why poor relief kept getting more expensive and a convenient justification for refusing to pay.

In 1817 the House of Commons appointed a commission of devout Malthusians to inquire into the system that at the time provided subsistence to thousands of Britons. Their report presented a smattering of unanalyzed data before launching into a theoretical exposition of population surplus. Condemning unemployment relief as "perpetually encouraging and increasing the amount of misery it was designed to alleviate," the commission recommended the adoption of the workhouse system and the immediate elimination of the allowances paid to families that increased in proportion to the number of children.[15]

Just about everyone concerned with the issue of Poor Law reform capitulated. Even the Evangelical Christians, who had labored for decades to put the existing welfare system in place, embraced the report's conclusions.[16] The most important Evangelical newspaper admitted that the Poor Law "has rather burdened us by its weight than supported us by its strength," and hoped that "some strong hand will, in time, bring it to the ground."[17] The Evangelicals in Bath, who had been engaged in welfare work for twenty years, abandoned charitable activities in the 1820s in favor of a Malthusian crusade to extirpate pauperism, arguing that the Poor Laws only added to "the sum of poverty and misery."[18] Malthus's twin victories—over Christian humanitarianism, on the one hand, and the revolutionaries' "right to subsistence," on the other—marked the complete ascendancy of utility. The specter of hunger and poverty was to be tolerated because to alleviate it with money would only make things worse in the long run. While the Evangelicals had argued for poor relief on the basis on Christian principles, the emphasis was now squarely on consequences. The debate about British social and political reform was won for utilitarianism.

As a prophet of utility over sentiment, Malthus went far beyond Bentham. His analysis was grounded in the starkest animal predicament of our species. He defined unemployment and pauperism as biological necessities. Even though the Malthusian solution would prove, in the end, to be no panacea, the questions he raised continued to reverberate. Eventually his analysis of procreation, hunger, survival, and extinction spread beyond social reform into the sciences of life. As a result, all of the ideas about progress and regress, happiness and misery, perfectibility and savagery that had proved so explosive during the French Revolution resurfaced, half a century later, in theories of biological evolution. These, in turn, raised questions about the evolutionary significance of pain and pleasure.

EXPERIMENTS WITH LAUGHING GAS

In a scathing review of the 1803 edition of Malthus's *Essay*, the poet Robert Southey cried: "No wonder that Mr. Malthus should be a fashionable philosopher! He writes advice to the poor for the rich to read."[19] Southey belonged to an unruly cabal of romantic revolutionaries trying to keep aloft Joseph Priestley's dream of political and scientific progress. For these men, Malthus's recommendations were not only heartless but out of step with ever-increasing human control of nature. For Priestley's acolytes, the world was clearly on the brink of a momentous change. Human ingenuity was beginning to liberate the powers of electricity, air, water, and steam. Soon these tools would be harnessed to ameliorate the miseries of embodied existence—to "destroy our pains and increase our pleasures."[20] In an age of scientific progress, it made no sense to talk of gloomy necessity or to insist, as Malthus had done, that societal institutions were "mere feathers that float on the surface" of the murky stream of animal and human life.[21]

Southey's circle included the physician Edward Jenner, who altered the Malthusian predicament forever with his 1798 pamphlet *Vaccination against Smallpox*. Jenner took the observation that milkmaids who caught the cowpox on their hands rarely got smallpox and turned it into a controllable, relatively safe method of protection against an ancient scourge. At the end of his pamphlet describing the technique, he predicted that his "vaccination"—from the Latin word for cow—would enable mankind to eradicate smallpox from the face of the earth; 180 years later, in 1978, his promise was fulfilled.

Malthus's essay on the inevitability of pestilence and famine, in other words, was published the same year as the first medical treatise to have a significant and lasting effect on mortality statistics. With the simultaneous publication of Bentham's "Houses of Industry" scheme, Malthus's population principle, and Jenner's treatise on vaccination, the fundamentals of the British debate about material progress were set. Can we transcend our animal predicament? Can reason free us from the pains and limits that hobble other species? Is poverty eradicable?

For Malthus the answer to each of these questions lay somewhere between "no" and "unlikely." Bentham might have replied with a cautious affirmative, but only insofar as Jeremy Bentham Esq. was anointed philosopher king of the globe (an opportunity humanity seemed inexplicably willing to forgo). For Jenner and his comrades, the response would have been a resounding "yes." Full of a sense of their own intellectual power, these men inhabited a world that was malleable and full of promise. For them, left-wing politics and experimental science together could be fashioned into a medicine for all of society's ills.

After Priestley's 1794 departure for America, his disciples began to tinker around with the new gases he had discovered, trying to develop "pneumatic" therapies for respiratory diseases. Foremost among the experimentalists was Humphry Davy, a woodworker's son from Cornwall, possessed of an autodidact brilliance that could turn ordinary seaweed into scientific equipment. As a child, Davy had made fireworks to his own recipe and fashioned ingots of tin by melting down fragments with a candle in a hollowed-out turnip. By the time he was in his teens, he had contrived to make an air pump out of a rusted enema syringe that had washed up on the beach and had conducted a series of experiments on the composition of the air inside seaweed bladders. Apprenticed to an apothecary, he was discovered swinging on a gate in Penzance by an associate of Priestley's, who questioned the sixteen-year-old about chemistry, found that he was a scientific prodigy, and brought him to Bristol to work in a proper laboratory.[22]

Years earlier, Priestley had synthesized a new gas by heating iron filings dampened with nitric acid. Davy began to experiment with this substance, developing more efficient ways of collecting it and seeing what effects it had on small mammals. Then he began to self-experiment. On April 17, 1799, he became the first person to inhale a psychoactive dose of pure nitrous oxide. He noticed a curious sweet taste and then "a highly pleasurable thrilling in the chest and extremities." After a few minutes of inhaling, he dropped the oiled silk bag in which he had collected the gas, jumped to his feet, and started "shouting, leaping, running" like one "excited by a piece of joyful and unlooked-for news."[23] He spent the rest of the year probing the significance of this happy bulletin from his nervous system. The published account of his investigations appeared in 1800 as *Researches, Chemical and Philosophical; Chiefly Concerning Nitrous Oxide.* One chapter consisted of the first-person reports of his circle— including Southey, Samuel Taylor Coleridge, and Peter Roget of *Roget's Thesaurus* fame—detailing their experiences of laughing gas.

Davy's *Researches* distills the optimistic essence of liberal Epicure-anism at the threshold of the nineteenth century. Even as the severed heads piled up in Paris, Priestley's apostles had defiantly continued to champion the Jacobin cause. When the Revolution began to fall apart, they turned to scientific medicine for solutions to social problems. The Pneumatic Institute in Bristol, where the nitrous oxide experiments took place, was supposed to be the prototype of a better future. Infected with Davy's contagious enthusiasm for the gas, Priestley's heirs got high, one at a time. Most found themselves involuntarily capering around in front of their friends, fizzing with pleasure, and laughing hysterically. Together the more radical members of the group plotted a new utopia of unbounded pneumatic pleasure: communal farming, fresh food, fresh air, sexual equality, libertarian childrearing, and political freedom. This proto-hippie recipe for the good life went by the absurd name of "pantisocracy," and Priestley was supposed to be preparing the ground for a practical demonstration in Pennsylvania's Susquehanna Valley.

By virtue of his youth, Davy himself was mostly untouched by his circle's larger ambitions for social change. He had been four years old when the Bastille was stormed, and he was simply too young to have been raised up and cast down by the hopes and disappointments of revolution. The autobiographical sections of his *Researches* display more appetite for chemical pleasure than political change. Gradually increasing the fre-quency and intensity of his sessions with the gas, Davy ended up availing himself of the "breathing-box," which had been designed for Pneumatic Institute patients too weak to hold the bag to their lips. The equipage essentially consisted of a sedan chair in an airtight paneled chest, and Davy settled inside, naked to the waist, while his colleague Dr. Kinglake

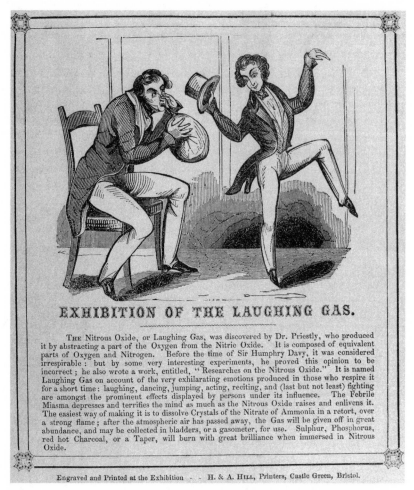

EXHIBITION OF THE LAUGHING GAS.

THE Nitrous Oxide, or Laughing Gas, was discovered by Dr. Priestly, who produced it by abstracting a part of the Oxygen from the Nitric Oxide. It is composed of equivalent parts of Oxygen and Nitrogen. Before the time of Sir Humphry Davy, it was considered irrespirable: but by some very interesting experiments, he proved this opinion to be incorrect; he also wrote a work, entitled, "Researches on the Nitrous Oxide." It is named Laughing Gas on account of the very exhilarating emotions produced in those who respire it for a short time: laughing, dancing, jumping, acting, reciting, and (last but not least) fighting are amongst the prominent effects displayed by persons under its influence. The Febrile Miasma depresses and terrifies the mind as much as the Nitrous Oxide raises and enlivens it. The easiest way of making it is to dissolve Crystals of the Nitrate of Ammonia in a retort, over a strong flame; after the atmospheric air has passed away, the Gas will be given off in great abundance, and may be collected in bladders, or a gasometer, for use. Sulphur, Phosphorus, red hot Charcoal, or a Taper, will burn with great brilliance when immersed in Nitrous Oxide.

Engraved and Printed at the Exhibition - - H. & A. HILL, Printers, Castle Green, Bristol.

Figure 6. "Exhibition of the Laughing Gas," wood engraving, circa 1840. The print shows two young men at a laughing gas demonstration, one inhaling the substance from a silk bag and the other capering about, illustrating the effects of nitrous oxide described by Humphry Davy and his circle. The engraving was printed by H. & A. Hill in Bristol, the city where the experiments took place, and betrays a certain civic pride in Davy's legacy. Brimful of faith in technological and medical progress, the laughing gas experiments represent the polar opposite of the Malthusian pessimism that was gaining ground at the time. Courtesy of the Wellcome Library, London.

pumped in eighty quarts of nitrous oxide. The last blast of the gas gave Davy "a great disposition to laugh," while "luminous points" floated in front of his eyes.

After an hour and a quarter inside, he staggered out of the box and promptly inhaled another twenty quarts of "unmingled nitrous oxide" from the proffered bag:

> A thrilling extending from the chest to the extremities was almost immediately produced. I felt a sense of tangible extension highly pleasurable in every limb; my visible impressions were dazzling and apparently magnified, I heard distinctly every sound in the room and was perfectly aware of my situation. By degrees as the pleasurable sensations increased, I lost all connection with external things; trains of vivid visible images rapidly passed through my mind and were connected with words in such a manner, as to produce perceptions perfectly novel. I existed in a world of newly connected and newly modified ideas. I theorized; I imagined that I made discoveries. When I was awakened from this semi-delirious trance by Dr. Kinglake, who took the bag from my mouth, indignation and pride were the first feelings produced by the sight of the persons about me. My emotions were enthusiastic and sublime; and for a minute I walked round the room perfectly regardless of what was said to me. As I recovered my former state of mind, I felt an inclination to communicate the discoveries I had made during the experiment. I endeavoured to recall the ideas, they were feeble and indistinct; one collection of terms, however, presented itself: and with the most intense belief and prophetic manner, I exclaimed to Dr. Kinglake, *"Nothing exists but thoughts!—the universe is composed of impressions, ideas, pleasures and pains!"*[24]

Davy's "indignation and pride" at being summoned back to reality perfectly captures that ludicrous combination of loss of motor control with increased self-esteem that sometimes comes with chemical intoxication. As for the content of his stoned prophecy, there it is—the ineffable insight from beyond the doors of perception, in all its pristine vacuity.

Davy survived the inevitable jokes about flatulence and hot air that followed the publication of the book and went on to have a dazzling career as the greatest British chemist of his time. Another member of the laughing gas circle was not so fortunate. The principal patron of the

Pneumatic Institute was Tom Wedgwood, consumptive heir to a great industrial fortune and himself a superb experimentalist. By the time his lungs failed him completely at the age of twenty-two, he had published two papers on heat and light with the Royal Society, anticipating a general theory of incandescence that would not be formulated for another fifty years. Confined to his bed and resigned to a life of invalidism, in 1793 Wedgwood turned his full attention to the task of introspecting about pain and pleasure.

Wedgwood's notes exemplify the desperation of Priestley's disciples as they tried to keep their ideals intact in an increasingly conservative and repressive Britain. His portrait of humanity was almost as unsparing as Hobbes's: "Never was there yet a human being who had the honesty to declare all the sources of his pleasures. . . . What greater treat than a house on fire?" But his ultimate vision, informed by his own urgent need for pain relief, was still madly utopian. "Five minutes in a Man's life may contain an intensity of enjoyment equal to that of as many months in any other animal," he declared. To maximize happiness, "Man should multiply on the face of the Globe to the exclusion, if it be necessary, of every other species."[25] Wedgwood's answer to the "population question" demonstrates the commitment to the uniqueness of human life that differentiated his circle from the devotees of Bentham and Malthus. We may be animals, but we are rational animals, they urged. We have minds. We have rights and duties, culture and civilization. Our pleasures are of a different order of magnitude than those of any other material being. If an increase in happiness is the goal, then human life, the more the better, is an end in itself.

Wedgwood's wild aspirations for human happiness were at the same time bulletins from the tragic edge of medical failure. His report of his experience with nitrous oxide in Davy's book contains no trace of the once-bright hope that laughing gas might work as a treatment for his consumption. He describes inhaling the gas, until he became "as it were entranced," at which point he threw the bag away and breathed furiously with an open mouth while holding his nose with his left hand, "having no power to take it away though aware of the ridiculousness of my situation."[26] Even in the midst of this hilarity, the collective inability of the pneumatic revelers to come up with anything in the way of a treatment for Wedgwood's tuberculosis is a sober reminder of the prematurity of their plans for humanity. He died, delirious with pain and laudanum, in 1805, after a last heroic effort in the laboratory, a collaboration with Davy on "sun pictures" produced according to the basic principles of photography. He had invented all of the steps to make a photograph, except a way to fix the image, so that the silver nitrate did not darken to uniform blackness upon exposure to ordinary light. Priestley and his "sons of genius" had caught a tantalizing glimpse of a technoscientific future, without having the faintest idea of how long it would take to get there.

THE SPRINGS OF ACTION

Meanwhile, Bentham was continuing to develop his own brand of political medicine. In 1802 his Swiss editor, Étienne Dumont, published his *Traités de Legislation* in three volumes. This tract made him famous. As Dumont remarked in the preface, Bentham "discarded all dogmatic words" and "rejected all terms that do not express some sensation of pleasure or of pain." Sales were brisk in Russia, and word spread to Spain, Italy, Greece, Portugal, and South America. Back on his own turf, things were not quite as adulatory: as one commentator observed, British critics had been amusing themselves "like the *valet-de-chambre* of a hero, with his foibles and peculiarities at home."[27]

In this new work, Bentham elaborated on the idea of the legislator as a medical practitioner. Here is a passage from the English translation of the *Traités de Legislation*, worth quoting in full for his analogy between political reform and therapeutic intervention:

> Pathology is a term used in medicine. It has not hitherto been employed in morals, but it is equally necessary there. When thus applied, moral *pathology* would consist in the knowledge of the feelings, affections, and passions, and their effects upon happiness. Legislation, which has hitherto been founded principally upon the quicksands of instinct and prejudice, ought at length to be placed upon the unmoveable base of feelings and experience: a moral thermometer is required, which should exhibit every degree of happiness or suffering. . . . Medicine is founded upon the axioms of physical pathology; morals are the medicine of the soul; legislation is the practical branch; it ought, therefore, to be founded upon the axioms of mental pathology.[28]

These sentiments were more than metaphorical: in this passage Bentham established the framework that would unite medicine, utilitarianism, and politics for the next hundred and fifty years. It was through its effects on the nervous system—the pains of bad laws and the pleasures of good ones—that political reform qualified as practical medicine.

Just as Bentham was enjoying his first taste of worldwide fame, an artist-physician named Charles Bell was conducting a series of experiments on rabbits, attempting to track what he called the "springs of action" in the nerve fibers. After anesthetizing the creatures with a judicious knock on the head, Bell opened them up to investigate the two separate bundles of fibers that issue from the spinal cords of vertebrates. He found that pinching one bundle of nerves produced convulsions in unconscious rabbits, whereas the other did not. He concluded that *movement* was controlled exclusively by the first nerve bundle. But Bell had to content himself with only half a discovery. Too tenderhearted to conduct surgery on conscious creatures, he never ascertained what the second bundle did.[29]

Credit for completing the investigation goes to the legendarily callous French physiologist François Magendie. Famous for an experiment in London in which he nailed the paws and ears of a living, wide-awake dog to the table and left it there overnight, Magendie took up the task of replicating Bell's experiments on conscious animals. He confirmed Bell's finding that stimulating one bundle caused the animals to twitch or convulse. And he discovered that stimulating the other bundle of fibers made the creatures cry out in pain. Later experiments found that cutting the first bundle paralyzed animals; cutting the second anesthetized them. It appeared that one lot of fibers conveyed sensation from the body to the brain; the other carried movement from the brain to the body.

In 1815 Bentham published his *Table of the Springs of Action*. He used the same mechanistic vocabulary as Bell—actions are caused by "springs"—but in this case the springs were human pains and pleasures. The *Table* is impressively earthy. The first square concerns the "PLEASURES AND PAINS—*of the* TASTE—*the* PALATE—*the Alimentary Canal—of* INTOXICATION." The second deals with the pleasures and pains of sex, and the third with sensual pleasures more generally. Only after rummaging around in our animal nature does Bentham turn his attention to such human questions as money, power, religion, and work. The

No. I. PLEASURES AND PAINS,
—*of the* TASTE—*the* PALATE—*the alimentary canal—of* INTOXICATION.

Corresponding Interest,
Interest of the PALATE—Interest of the BOTTLE.

Corresponding *MOTIVES*—with NAMES,

—I. NEUTRAL: viz.	—II. EULOGISTIC: viz.	—III. DYSLOGISTIC: viz.	
1. Hunger.	Proper, none.		
		1. Gluttony.	vouring, gormandiz-
2. Need of food.	*Improper.*	2. Gulosity.	ing, guttling. &c.
3. Want of food.	1. Love of the pleasures	3. Voracity.	10. Drunkenness.
4. Desire of food.	of the social board—	4. Voraciousness.	11. Ebriety.
5. Fear of hunger.	of the social bowl, or	5. Greediness.	12. Intoxication.
	glass—of good cheer	6. Ravenousness.	13. Sottishness.
7. Thirst.	—of good living—of	7. Liquorishness.	
8. Drought.	the good goddess—of	8. Daintiness.	Love &c. (*as per Col.3.*)
	the jolly god, &c.		of &c. drink, liquor—
9. Need, want, desire		9. Love, appetite, crav-	drinking, tippling,
—of the means of		ing, hankering, pro-	toping, boosing, guz-
quenching, relieving,		pensity, eagerness,	zling, swilling, soak-
abating, &c. thirst.		passion, rage—of, for,	ing, sotting, carousing
		to, and after—cram-	—junketing, revell-
10. Inanition.		ming, stuffing, de-	ing, &c.

No. II. PLEASURES AND PAINS,
—*of the sexual appetite, or of the sixth Sense.*

Corresponding Interest,
SEXUAL INTEREST.

Corresponding *MOTIVES*—with NAMES,

—I. NEUTRAL: viz.	—II. EULOGISTIC: viz.	—III. DYSLOGISTIC: viz.	
Single-worded, none.	None.		
		1. Venery.	6. Libidinousness.
Many-worded,		2. Lust.	7. Lecherousness.
Sexual desire.		3. Lechery.	8. Salacity.
		4. Lewdness.	9. Salaciousness.
		5. Lustfulness.	
			10. Venereal desire.

No. III. PLEASURES AND PAINS,
—*of* SENSE, *or of the* SENSES: *viz. generically or collectively considered.*

Corresponding Interest,
Interest of SENSE—of the senses:—SENSUAL INTEREST.

Corresponding *MOTIVES*—with NAMES,

—I. NEUTRAL: viz.	—II. EULOGISTIC: viz.	—III. DYSLOGISTIC: viz.	
Single-worded, none.	None.		
		1. Sensuality.	8. Love, appetite, craving,
Many-worded,		2. Luxury.	&c. (*as per No. I. Col*
Physical want, need,		3. Carnality.	3.) of, for, to, and af-
exigency, necessity—		4. Debauchery.	ter—sensual pleasure,
desire, appetite.		5. Intemperance.	enjoyment, gratifica-
		6. Luxuriousness.	tion, indulgence, &c.
		7. Voluptuousness.	See note (*b*). *Synonyms*
			to pleasure.

Figure 7. Jeremy Bentham's "Table of the Springs of Action," 1815. This detail shows the first three squares of Bentham's table, dealing with the pains and pleasures of eating, sex, and sensuality in general. One of his aims was to develop a rigorously neutral language for these phenomena, resulting in his coining of the phrase "sexual desire." His alliterative list of synonyms for drunkenness shows Bentham enjoying the music of the evaluative language that he professed to despise: "tippling, toping, boosing, guzzling, swilling, soaking, sotting, carousing—junketing, reveling, &c."

penultimate square then returns to brute sensation with a thud of final-
ity: the "PAINS—*of* DEATH *and* BODILY *pains in general.*"[30]

Bell and Magendie's idea of the "sensory motor system" was stun-
ning in its simplicity and far-reaching in its explanatory power. Often
compared to William Harvey's discovery of the circulation of the blood,
this mechanism of inputs and outputs was beautifully compatible with a
psychology centered on pleasure and pain. Sensation travels *to* the brain
along one pathway; instructions for movement travel *from* the brain
along another. When the sensation is pleasurable, the muscles move to
prolong or repeat it. When the incoming message is painful, the outgoing
instruction tells the flesh to flinch away. Hobbes's mechanistic geometry
of human behavior—toward delight and away from misery—had found
its anatomical basis. As we will see in the next chapter, it was when sensory motor physiology was joined to pleasure-pain psychology—when
Bell's springs of action were united with Bentham's—that utilitarianism
was transformed into an experimental science of behavior.

THE HISTORY OF INDIA

In 1808 Bentham recruited his most devoted, influential, and energetic
disciple, the journalist, historian, and imperial administrator James Mill.
Like Malthus—professor of political economy at the East India College at
Haileybury—Mill had global ambitions for his utilitarian reform agenda.
Bentham disapproved of imperialism on principle, but Mill had no
qualms about imposing his agenda on whatever distant lands were lucky
enough to come under the sway of British civilization. Much of the colonial reach of Victorian utilitarianism can be attributed to his personal
promotion of the cause.

James Mill was born James Milne in Scotland in 1773. His father
was a shoemaker, as was his father before him, but his mother was socially ambitious. It was she who changed the family name from Scottish-
sounding "Milne" to the English-sounding "Mill." Finding that her eldest
son was gifted, she freed him from household chores and from toiling in
his father's trade in order that he might pursue his studies and become
a gentleman. The boy went to the parish school and then to a grammar
school. There his talents came to the attention of Sir John and Lady Jane
Stuart, members of the local gentry who had established a fund for the

education of boys for the ministry. Under their sponsorship, Mill was sent to Edinburgh University. By his midtwenties he had become licensed as a Scottish Presbyterian preacher.[31]

The spiritual aspects of Mill's education did not stick. Although it is not clear precisely when he abandoned the Presbyterian Church, we do know that in 1802 he ventured to London with the resolution to make his living by his pen rather than by sermonizing. He wrote for the *Anti-Jacobin Review* and signed up for the volunteer army in case of an invasion by Bonaparte. Years of journalistic and editorial drudgery followed. In 1806 he embarked on a vast undertaking: to write a history of India, from the earliest times to the present. The same year, his first child was born, a son whom he named John Stuart after his aristocratic benefactor.

Two years later, Mill made the acquaintance of Jeremy Bentham, then sixty years old. The relationship was that of master and disciple. At first Mill would walk from Pentonville to Westminster to dine with his hero; thereafter, Bentham seems to have arranged for Mill and his ever-growing family to rent a series of London houses nearer to his own residence, and to spend their summers at whatever country house he happened to be inhabiting.

In 1814 Bentham became the tenant of Forde Abbey, a beautiful Tudor house built on the remains of a medieval monastery in Dorset. The buildings included a private chapel and cloisters hung with tapestries. The grounds boasted a quarter-mile-long avenue of trees, a lake, and a deer park. Mill spent the summers in residence, and he and Bentham would sit working on their respective projects at either end of one of the lofty rooms.[32] Their alliance marked the founding of a school of followers, some of them members of Parliament, who dubbed themselves "philosophical radicals" in celebration of the rigor and the purity of their political creed. Subjecting all ideas about political reform to the test of utility, they debated how far the extension of the franchise should reach in order to maximize happiness, and what to do about the problem of poverty.

James Mill thought his *History of India* would take him three or four years; in the event, it took up nearly twelve years of his life. The first edition was published in 1817. Most of the merchants, adventurers, and soldiers who had traveled or lived in India in the eighteenth century professed great admiration for its ancient civilization. Mill, by contrast, assumed a fixed scale of cultural progress and placed India on a low step.

Previous commentators, he explained, had not the advantage of a science of social progress to guide them in their estimation of Indian society. For Mill, the standard of utility provided him with an Olympian vantage point from which to judge all the past, present, and future deeds of men: "Exactly in proportion as *Utility* is the object of every pursuit, may we regard a nation as civilized," he announced.[33]

Mill was almost as critical of the British regime in India as he was of its Mogul predecessor, but despite this unsparing verdict he was appointed "Assistant to the Examiner of Indian Correspondence" in 1818, at the handsome salary of eight hundred pounds a year. He was to work at India House for the rest of his life, installing his eldest son in the same office on Leadenhall Street at the age of seventeen. From their standing desks, father and son administered the empire (neither man ever set foot in India), promoted Benthamite social and political reforms, and produced a body of philosophical work distinguished by its application of first principles to the practical problems of the nineteenth-century metropolis.

And problems there were aplenty. The hungry years after the peace of 1815 were tumultuous ones, marked by riots, demonstrations, uprisings, and a ceaseless flow of radical pamphlets denouncing the ruling class. The threat of revolution made reform more appealing, and the philosophical radicals were suddenly called upon to put their principles into practice. In 1817 Bentham published a "Catechism of Parliamentary Reform," scandalizing the Whigs by advocating universal suffrage ("females might even be admitted") and annual elections.[34] Written in the form of a dialogue, the "Catechism" was both too abstract and too radical to be absorbed into a party platform, and it fell to Mill to press the cause of democracy in a more conservative and accessible form. The time was ripe. In 1819 the Peterloo Massacre, at which fifteen people were killed when the cavalry charged a demonstration in Manchester, marked the deepening polarization between the conservative and radical forces in the country and provoked a new round of motions for parliamentary reform. In 1820 Mill's friends and admirers persuaded him to republish an article on government, originally written for the *Encyclopaedia Britannica*.

The essay opens with the Benthamite premise that the purpose of government is the greatest happiness of the greatest number; it proceeds via the Hobbesian axiom that all men will seek to enslave one another in the absence of any check to their actions; and it ends with the liberal conclusion that representative government is the best way to minimize the harmful results of human selfishness. The coda was that men over forty with property suffice to represent the interests of the whole nation. With the publication of this article, these diluted Benthamite principles—females, for example, were *not* admitted to the franchise in Mill's scheme—became the manifesto of the philosophical radicals in Parliament.

Mill had opened his essay on government with the suggestion that "the whole science of human nature must be explored." In 1822 he duly began work on his *Analysis of the Phenomena of the Human Mind*. The last quarter of this book is given over to a consideration of pain and pleasure, which "hold a great share in composing the springs of human action."[35] Importantly, for Mill, pains and pleasures do not remain in consciousness as raw sensations. Rather, they quickly direct attention away from themselves to the causes that give rise to them, "that we may prevent, or remove it, if the sensation is painful; provide, or detain it, if the sensation is pleasurable."[36]

Mill's last few years were busy ones, taken up with various projects for the implementation of utilitarian reforms at home and abroad.[37] By the 1830s some of his more sweeping views on the Indian administration were being put into practice, especially regarding the enclosure of private property, an ambitious attempt to impose a whole system of law and government on a distant land. The idea was to make a Domesday survey of land rights across the entire subcontinent and then erect a system of law to adjudicate property claims, whereupon "crimes and litigation would soon cease, and human conduct would be permanently canalized into beneficial or harmless courses."[38] For Mill, even more than for Bentham, it was private property that promoted the cause of happiness.[39]

Twenty-five years Bentham's junior, Mill was far more conservative than his mentor. Bentham had initially been sympathetic to the French Revolution; Mill started his journalistic career writing for the *Anti-Jacobin Review*. Bentham believed that individuals were the best guardians of their own interests; Mill believed that Indians were constitutionally incapable of promoting their own happiness. Bentham firmly subscribed to a universal human nature; Mill arranged nations along a scale of progress and regress, with British utilitarians at the apex.

THE MAN BEHIND MISS GRADGRIND

While Bentham and James Mill worked in the tapestry-hung rooms of
Forde Abbey, they raised Mill's gifted oldest son in the utilitarian creed.
Starting at an age that now seems scandalously young, John Stuart Mill
was subjected by his father to a regime of homeschooling that has become legendary for its rigor and intensity.[40] He started Greek and mathematics at three, Latin at eight. He recorded that his father "demanded
of me not only the utmost that I could do, but much that I could by no
possibility have done."[41]

James Mill's goal was that his son should inherit the Benthamite
mantle and become the greatest utilitarian reformer in history. When
the boy was a prodigiously erudite six-year-old, Bentham promised to
take over his education in the event of his father's death and "by whipping
or otherwise, do what soever . . . necessary and proper, for teaching him
to make all proper distinctions, such as between the Devil and the Holy
Ghost, and how to make Codes and Encyclopedias." James Mill replied
with the pious hope that "we may perhaps leave him a successor worthy
of both of us."[42]

At the age of fifteen, Mill junior seemed to be fulfilling the two reformers' fondest ambitions when he formed "a little society, to be composed of young men . . . acknowledging Utility as their standard in ethics
and politics . . . and meeting once a fortnight to read essays and discuss
questions."[43] A scant two years later, his father fixed him up with a career
in the Office of the Examiner of Indian Correspondence "immediately
under himself."[44] Father and son conducted the business of empire together and collaborated on articles for the new journal of the philosophical radicals, the *Westminster Review*. James Mill was the elder statesman
of the utilitarians; John Stuart Mill was their young leader. Father and
son stood shoulder to shoulder, sharing the yoke of social improvement,
at home and in the empire.

In 1826, when Mill junior was twenty years old, he suffered a bout of
depression serious enough that he contemplated suicide. Executing his
usual round of writing and debating, he "went on with them mechani-
cally, by the mere force of habit. I had been so drilled in a certain sort of
mental exercise, that I could still carry it on when all the spirit had gone
out of it."[45] He discovered that there was no one in his life "to whom I

had any hope of making my condition intelligible." Even his distress itself seemed meaningless, and he judged it to have "nothing in it to attract sympathy."[46]

Mill later wrote that "my case was by no means so peculiar as I fancied it." But, he added, "the idiosyncrasies of my education had given to the general phenomenon a special character."[47] He felt that he had been so relentlessly trained in the habits of utilitarian analysis that all natural feeling had dried up in him, leaving him "no delight in virtue or the general good, but also just as little in anything else."[48] His father believed the purpose of education was to form "associations of pleasure with all things beneficial to the great whole and of pain with all things hurtful to it." The hapless object of this pedagogical experiment could only declare it a failure.[49] All it had done was to make him numb to all feeling. Almost exactly halfway between Hobbes's *Leviathan* and B. F. Skinner's *Beyond Freedom and Dignity*, we have arrived at the first emergence of an applied utilitarian psychology, dedicated to the creation, in Mill's bitter words, of "a Benthamite, a mere reasoning machine,"[50] whose desires and aversions had been manipulated since infancy for the greater good.

The depression eventually lifted, but it would leave a profound mark on Mill's thought. Searching for a way out of his emotional impasse, he read widely in the writings of anti-utilitarian reactionaries, romantics, and revolutionaries, such as Carlyle, Goethe, and Wordsworth. In 1830 his loneliness was finally overcome when he met Harriet Taylor, at her husband's table. She was twenty-three, he was twenty-five. They carried on a much-gossiped-about intimate friendship for twenty years, until Mr. Taylor's death freed them to marry. She died of lung failure a short seven and a half years later, leaving Mill distraught with grief.

Mill's collaboration with Taylor resulted in his most famous work, *On Liberty*. "The object of this essay," he declared, "is to assert one very simple principle . . . That the only purpose for which power can be rightfully exercised over any member of a civilized community, against his will, is to prevent harm to others."[51] Should people be tempted to interfere with us for our own good, Mill insisted that they must desist, "even though they should think our conduct foolish, perverse or wrong."[52] Two years later, Mill published his other much-anthologized essay, *Utilitarianism*. Appealing to our "sense of dignity" to explain why "it is better to be a human dissatisfied than a pig satisfied," he signaled his distance from the Hobbesian psychology of his father, who had dismissed "dignity" in his *Analysis of the Phenomena of the Human Mind* as nothing more than a symptom of Man's incurable egotism.[53]

Mill's Benthamite upbringing was satirized in Charles Dickens's 1854 novel *Hard Times*, set in a fictional northern mill town. The heroine of the book is Louisa, favorite child of the utilitarian pedagogue Thomas Gradgrind. Louisa's fine and selfless nature has been crushed by the rigors of her education. In one scene, she asks her father, "What do *I* know . . . of tastes and fancies; of aspirations and affections; of all that part of my nature in which such light things might have been nourished? What escape have I had from problems that could be demonstrated, and realities that could be grasped?"[54] In the novel, Thomas Gradgrind eventually repents of his treatment of his daughter, and she frees herself from her spiritual bondage to utilitarianism. Similarly, John Stuart Mill liberated himself from both his father and Bentham, and embraced a Romantic credo of self-development.

Throughout his whole life, however, the younger Mill remained steadfast in his loyalty to the most controversial of all utilitarian fathers, the Reverend Malthus. "Malthus's population principle was quite as much a banner, and point of union among us," he recalled of his band of young utilitarians, "as any opinion specially belonging to Bentham."[55] At the age of seventeen, he came across a dead baby in a park and started distributing contraception advice pamphlets in the poorer districts of London, for which he was arrested and spent several days in jail.[56] Two years later, when he delivered a speech on Malthus, he admitted that he felt "considerable embarrassment" at the prospect of defending views on which "Tory, Whig and Radical, however they may differ in other respects, agree in heaping opprobrium."[57] But Malthus's theory appeared to him "to rest upon evidence so clear and so incontrovertible, that to understand it, is to assent to it, and to assent to it once is to assent to it for ever."[58] And "assent to it for ever" he did, still asking, nearly twenty years later, why the middle classes should sacrifice a share of their wealth "merely that the country may contain a greater number of people, in as great poverty and as great liability to destitution as now?"[59] Trying to reconcile Malthus's stress on outcomes with the highest reaches of Continental romanticism, Mill's morality was riven by the same tensions between autonomy and utility, paternalism and liberty, the individual and the collective, body and soul, that trouble the edifice of twenty-first-century medical ethics.

THE ANATOMY ACT

In 1830 revolution swept France once again. On the domestic front, public agitation for parliamentary reform mounted. When a second re-form bill was voted down by the House of Lords in 1831, Bentham was acclaimed a hero by the demonstrators, who took to the streets to agitate for "the Bill, the whole Bill and nothing but the Bill." The violence and passion of the demonstrations shocked everyone. Nottingham Castle was burned down; in Bristol the Tory MP was assaulted; in Birmingham, a meeting of the newly formed Political Union resolved to pay no taxes un-til reform was passed. In December, Earl Grey, the Whig prime minister, hastily introduced a third bill, which passed the Commons with a big ma-jority. This version of the Great Reform Act abolished the so-called rotten boroughs—places with few if any inhabitants that were still represented in Parliament—as well as extending the franchise to all males with prop-erty worth ten pounds or more. Again the Lords rejected it. Eleven days of rioting followed, and Queen Adelaide voiced her conviction that she was going to share the fate of Marie Antoinette. Eventually William IV threatened the Lords with the creation of a majority of Whig peers, and they grudgingly allowed the Great Reform Act to go through.

On the same day as the passing of the Great Reform Act, a lesser-known piece of legislation sailed through Parliament in its tailwind. This was the Anatomy Act, the purest piece of Benthamite reasoning ever to make it into law.[60] The problem was the supply of dead bodies. Af-ter Napoleon's 1815 defeat, with travel now safe in Europe, and with a plentiful supply of cadavers available in the great teaching hospitals of Paris, students from all over the world began to flock to France to learn the arts of pathological anatomy. Everywhere else, medical schools suf-fered plummeting enrollments. The German Confederation responded with liberalization of laws regarding dissection, as did Vienna and most of Italy. In Britain, however, the supply remained severely limited. The 1752 "Murder Act" remained on the books, which gave judges the dis-cretion to include dissection as part of their sentencing for the heinous crime of homicide. The legal supply of dead bodies for medical education purposes fell hopelessly short of the demand, and the so-called resurrectionists—a shady bunch who made a living digging up corpses and selling them to anatomists—were kept in business.

In the 1820s a proposal began to circulate to adopt the Continental system of allowing medical students to dissect the bodies of any person buried at the public expense. In 1824 the *Westminster Review* reviewed the proposal in an article called "Use of the Dead to the Living," by a physician called Thomas Southwood Smith. Other commentators had protested that the Continental system would single out the bodies of the poor to be sacrificed on the altar of science.[61] In a deft move, Smith answered this objection, turning the proposal into a classic bit of utilitarian reasoning. Being buried at the public expense, he wrote, implied that no one had come forward to pay for a funeral. "It is not proposed to dispose in this manner of the bodies of all the poor: but only that portion of the poor who die unclaimed and without friends, and whose appropriation to this public service could therefore afford pain to no-one."[62] By identifying a population of people who left behind no relatives or companions to mourn them, Smith believed that he had whittled down the payment side of the utilitarian calculus to zero.

Hereafter, the utilitarian solution to the anatomy crisis in Britain would be that only "unclaimed" bodies be dissected. In 1828 a parliamentary committee composed of loyal Benthamites was appointed to ponder the question. The resulting report recommended that "the bodies of those . . . who die in workhouses, hospitals and other charitable institutions, should, if not claimed by next of kin within a certain time after death, be given up, under proper regulations, to the Anatomist." The objection that the poor would be unfairly singled out could now be summarily dismissed: "Where there are no relations to suffer distress, there can be no inequality of suffering, and consequently no unfairness shown to one class more than another."[63]

In one of his letters to the prime minister on the issue, Bentham had voiced his fear that before long a murder would be committed in the name of anatomy, and, indeed, in October 1828 the body of an elderly woman was discovered in Edinburgh, hidden in the bedstraw of the cobbler William Burke. It emerged that he and his confederate William Hare had committed sixteen murders in order to sell the bodies for dissection.

The two men had stumbled upon the procedure by accident, when an old man died in their boardinghouse and they decided to sell his body to cover his debts. It was so easy and so profitable that later, when another lodger fell ill, they eased him into a coma with whisky and smothered him. Another fifteen people—all vagrants—were lured to their deaths before the murders were discovered.

The publicity surrounding the Burke and Hare case made a change in the law inevitable, and in March 1829 the radical MP Henry Warburton submitted his "Bill for Preventing the Unlawful Disinterment of Human Bodies, and for Regulating Schools of Anatomy."[64] This first attempt at what would become the Anatomy Act failed under pressure from paternalistic aristocrats in Parliament, who asserted the rights of the poor to a decent burial. In the turmoil of parliamentary reform in 1831–1832, however, a second bill was put forward, and this one was successful, passing its final reading in the House of Commons in May 1832 before being dragged through a demoralized House of Lords two months later on the coattails of the Great Reform Act.

As the Anatomy Act was making it through the last stages of the legislative process, Bentham died. Three days later, on June 9, Thomas Southwood Smith dissected his corpse before an audience of his friends and disciples. As described by one of the attendees, Bentham's autopsy was a solemn consecration of secular science and utilitarian ethics. The room in the Anatomy School was circular, with no windows but a skylight. The body lay under the skylight, clad in a nightdress with only head and hands exposed. Crowding around were his friends and disciples, as well as a class of medical students and a few physicians. A thunderstorm arose just as Smith commenced the dissection, and the lightning playing over the face of the deceased, rendered his placid expression "almost vital":

> With the feelings which touch the heart in the contemplation of departed greatness, and in the presence of death, there mingled a sense of the power which that lifeless body seemed to be exercising in the conquest of prejudice for the public good, thus cooperating with the triumphs of the spirit by which it had been animated. It was a worthy close of the personal career of the great philosopher and philanthropist. Never did corpse of hero on the battle-field, with his martial cloak around him, or funeral obsequies chanted by stoled and mitred priests in Gothic aisles, excite

such emotions as the stern simplicity of that hour in which the principle of utility triumphed over the imagination and the heart.[65]

Bentham had died, but in giving over his lifeless body to science he continued to exert a power to conquer prejudice in the name of the public good. The Anatomy Act became law two months later, on August 1, 1832. With this victory, Benthamite justifications for the use of the bodies of the poor became the moral foundation for medical research, first in Britain, and eventually across the whole Anglophone world.

THE SWAN STREET RIOT

Smith's argument for the "Use of the Dead to the Living" was a compelling one: the negative utility resulting from the violation of the feelings of a few superstitious paupers would be vastly outweighed by the improvement in the health and survival prospects of the numberless thousands who might profit from their sacrifice. Then, as now, the potential payoff from medical research and education can make the asymmetry between the two sides of the utilitarian calculus appear almost infinite.

We do, however, have an answer to the utilitarian case, in the form of Ruth Richardson's 1987 *Death, Dissection and the Destitute*, the definitive history of the Anatomy Act. Richardson has uncovered a startling number of acts of popular resistance to the practices of anatomists, bodysnatchers, legislators, and hospital physicians, both before and after the passing of the act. She has brought together contemporary reports of dozens of separate events: angry demonstrations outside the courtrooms where bodysnatchers were being tried; riots at anatomy schools and cholera hospitals; disturbances outside Parliament during debates on the Act; the dissemination of pamphlets against the Act; mass-exhumations at church burial grounds after bodysnatchers were seen to be at work; the burning of effigies of doctors; demonstrations at funerals; and riots at the gallows. Most important, she has analyzed the official definition of "unclaimed" and shown that the majority of the bodies categorized in this way were, in fact, claimed by friends and relatives, who happened to be too poor to pay for their loved ones' funerals. Above all, her book makes clear how thin was the Anatomy Act's veneer of universalism.

Over and over again, the popular campaign against the Act pointed out how carefully the better-off protected their own remains while consigning pauper bodies to the dissection table. The Paisley Reform Society asked, "Surely it cannot be expected that the poor will be jeered out of their feelings of abhorrence for a system, which the rich, the wise, and the powerful take such extraordinary care to guard against?" In 1829, in response to an earlier version, the butchers and salesmen of Leadenhall Market, working men who labored only yards away from where the Mills administered the British Empire, submitted a petition suggesting that "the High Dignitaries of the Church, and all the Judges of the land, and all Generals and Colonels commanding regiments, all Admirals and Captains in commission, men whose duty and ambition, and profit it is to serve their country in life" should be the ones to "dedicate their bodies after death to promote the advancement of knowledge."[66] The same year the radical medical journal the *Lancet* published a letter signed "One of the Unclaimed" from a man whose house had been foreclosed and whose wife and daughters had died, and who now languished in a workhouse "without a single relative to notice me."[67] Benthamite plans for "unclaimed" bodies continued unchecked.

Simmering working-class anger about the Anatomy Act occasionally boiled over into riot. On Friday, August 30, 1832, less than a month after the Act had passed into law, a three-year-old child died at the Swan Street cholera hospital in Manchester. At the funeral, the child's grandfather, a weaver named John Hayes, noticed that the coffin had no name chalked upon the lid, and began to suspect that the body had been tampered with. He climbed down into the grave to try to force it open, but it was firmly screwed down and resisted his attempts to bash it loose with a rock. After the ceremony, the sexton provided him with some tools and together they managed to pry open the box. It turned out to contain only the headless trunk of his grandchild, the head having been replaced with a brick wrapped in shavings.[68]

Hayes closed the coffin and covered it over with earth. After getting the assurance of the local Catholic priest that the matter would be looked into, he took a party of family and neighbors to the burial grounds to show them the evidence. He opened the coffin once again and held up the child's mutilated body. The coffin and the corpse were seized from him and paraded through the streets of Manchester, until a large crowd had formed. About two thousand people went to the hospital, where they forced open the gates and rushed through the building hurling chairs, tables, and other furniture into the yard. A witness saw the old man Hayes standing with a brick in his hand, exclaiming, "Give me my child's head and I'll give you your brick."

The crowd broke the ground-floor windows, dragged out bedding and furniture and set it alight, smashed the dead van into pieces, and threw it on the fire. No windows were broken in the wards. Ten people were arrested, and three troops of the 15th Hussars rode up and dispersed the six hundred people still remaining in the vicinity of the hospital. By eight o'clock "tranquility was restored." The priest to whom Hayes had first appealed initiated a search of the hospital and found the child's head. This was reattached to the body by the surgeon. On Monday morning the Board of Health convened an emergency meeting to consider the circumstances surrounding the riot. The act of mutilation was blamed on a Mr. Oldham, who had been appointed apothecary at the hospital only twelve days before and who had by then left town. The chairman announced that "in order to manifest the desire of the board to relieve the distressed feelings of the parents, the body would be placed in a leaden coffin, and then be allowed to be interred in the catholic burial grounds."

We know about this episode from a short article in the *Manchester Guardian*. The report is a whitewash. At night, "after the tumult had subsided," it narrates, the priest, "desirous to alleviate the distress of the parents as much as possible," called upon the chief surgeon to assist him in a search of the hospital. A "fruitless search was made throughout the hospital, and at Mr. Oldham's lodgings. But after prosecuting the search for some time further . . . the two gentlemen had the satisfaction of recovering the object of their inquiries." If the head was not found in the hospital or at the apothecary's lodgings, then where was it found? The awkward gap in the narrative suggests that its location may have been embarrassing to the hospital authorities. The priest and the surgeon then engaged in the delicate business of recapitation: "Mr. Hearne wrapped his handkerchief round it and conveyed it to the town hall, where Mr. Lynch, in his presence, again attached it to the trunk by means of thread stitching."

On Monday evening the child's second funeral took place. A large crowd, "of whom three-fourths were females," gathered outside St. Patrick's Chapel. The funeral procession arrived, the gates of the chapel were opened, and "hundreds rushed in." After the ceremony the priest climbed the steps in front of the altar and addressed the crowd, laying out the findings of the Board of Health, exculpating the hospital doctors, and denying the rumors then circulating that he had once raised from the dead a person who had died of cholera. He also announced that an application had been made to the board to allow all Catholics who died of the disease to be interred in the graveyard of the chapel rather than the cholera burying ground.[69]

The body of the child had been made whole and consecrated, and the promise was made to unite the spiritual community of Roman Catholics in the graveyard of the chapel, instead of dividing the paupers from the better-off. With these symbolic acts and strategic concessions, the surgeon, the priest, and the members of the Board of Health repaired the social fabric torn apart by the research imperatives of anatomical medicine. What could not be forever buried by the second funeral of John Hayes's granddaughter, however, was an awareness of the way that utilitarian medicine—from the passing of the Anatomy Act until the human experimentation crisis of the 1970s—tended to exacerbate rather than ameliorate asymmetries of class and power, by singling out for sacrifice those people it deemed to have the least to lose.

THE POOR LAWS

One of the devoted disciples at Bentham's deathbed was his personal secretary, Edwin Chadwick, a thirty-two-year-old lawyer interested in public health and radical reform. Chadwick had come to Bentham's attention with an 1828 article, "Preventive Police," arguing, in good utilitarian fashion, that it was better to prevent crime than to punish it. Recognizing a potential new disciple, Bentham invited Chadwick to come live with him and help him complete his Constitutional Code. For a year, Chadwick lived at Queen Square, assisting the master with the preparation of his manuscripts, right up to the time of his death.

The 1830s had started inauspiciously with a series of agrarian riots and rick burnings, sparked by three years of bad harvests. Rural laborers set fire to farmers' houses and barns and smashed up threshing machines. Machine manufacturing, the enclosure of common lands, and the vagaries of weather and soil all meant rising numbers of people thrown on the mercy of the parish relief work, where they were set digging for flints, picking up dung for fertilizer, or put in harness like donkeys to pull cartloads of stone. With the surge of interest in reform of the Poor Laws, Bentham coached Chadwick in his cherished principles of contract management and panoptic surveillance.

In 1832 Chadwick was appointed to the Poor Law Commission, where he discovered that "the prevailing doctrine, founded on the theory of Malthus, was that the general cause of pauperism was the pressure of population on the means of subsistence."[70] Chadwick successfully counterproposed that pauperism should be addressed by making life in a workhouse more painful than the existence of an independent laborer, in accordance with Bentham's "Houses of Industry" proposal. In the end, the 1834 Poor Law Amendment Act combined Benthamite and Malthusian approaches. The poor were given bread instead of cash, in order to avoid the Malthusian trap of driving up the price of food. In accordance with the best Benthamite psychology, the new "workhouse test" assumed that anyone unwilling to live in one of these grim places must not be completely destitute. In the workhouses, families were separated by sex and age, and the diet of inmates regulated "as in no case to exceed in quantity and quality of food the ordinary diet of the able-bodied labourers living within the same district." While acknowledging that the forced separation of husband and wife could be regarded as cruel and un-Christian, supporters of the measure argued, in good Malthusian fashion, that this enforced celibacy would check pauperism and lead "the labouring classes back to habits of self-reliance."[71]

Opinion was fiercely divided. Charles Dickens set the tone for the bill's critics with his portrayal of the cruelties of the workhouse system in *Oliver Twist*. The *Times* of London regularly published horror stories about workhouse conditions, reporting on whippings and starvation, overcrowding and filth. On the other side, supporters of the measure extolled its miraculous levitation of pauper morale, its impeccably humanitarian execution, and its salutary effect on the public purse. The truth lay somewhere between these two poles: the workhouses were not, on the whole, places of cruelty, the commissioners were not particularly strict in the enforcement of the workhouse test, and a large percentage of the indigent continued to receive monetary relief. As one historian has noted, in "an age of harshness, drunkenness, vagrancy, cruelty, and suffering," the Poor Law commissioners were far from heartless Dickensian villains.

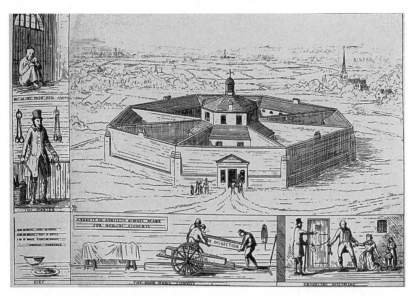

Figure 8. "Modern Poor House," from *Contrasts*, published in 1836 by the Roman Catholic architect Augustus Welby Pugin. Pugin's polemic decries the degeneration of British morals as evidenced in the decline of the Gothic style. For Pugin, modern building design deprived people of the elevating spectacle of an architecture that reached for heaven. His drawing of a "modern poor house" was a bitter visual attack on the corrupting, atheistic influence of the utilitarians. The building itself is a grim Panopticon in miniature, and the caption of the smaller image below it reads, "A variety of subjects always ready for medical students," a reference to the utilitarian Anatomy Act. The accuracy of Pugin's satire may be judged by comparing this image to the plan of the West Riding Pauper Lunatic Asylum in figure 9.

On the other hand, the same people "who wished workhouses physically comfortable, wished them psychologically unpleasant." Separated families could see each other only during mealtimes, when silence was enforced. There were rigid hours, early risings, prayers before meals. Punctuality and cleanliness were strictly enforced. The only outings permitted were to church. Most importantly for the present purposes, the law was based on a deeply "pessimistic view of the character of the poor."[72] The argument for the workhouse was that monetary relief payments would seduce the poor to pauperism, an understanding of human motivation reduced to the barest utility, in which indolence was always preferable to work.

Whatever were the rights and wrongs of the 1834 Poor Law, it represented the triumph of Malthusian and Benthamite naturalism, and signaled the complete emergence of a utilitarian theory of human nature. This view crystallized around the conviction that the natural, animal state of man was revealed at the hungry edge of the industrial economy. But the pauper's predicament was not brutish. The poor were not animals. They were men, women, and children. The laws of enclosure and industrial technologies that had dismantled the village economy and set labor "free" were not natural. Urbanization, laissez faire, and fossil fuels had produced a new sort of life—one that might, with an apology to Hobbes, be characterized as nasty, British, and short.

4

The Monkey in the Panopticon

MORAL PHYSIOLOGY

In the summer of 1831 the movement for parliamentary reform threatened to tip over into outright revolution. Scotland was a hotbed of political agitation, and in the city of Aberdeen thousands of Scottish loom weavers, blacksmiths, shopkeepers, merchants, and miners marched repeatedly along the ancient golf links at the beach to protest their corrupt government in London. Streaming alongside them, in processions of their own, were hundreds of children and teenagers, already working in the weaving sheds, factories, and mills of the Granite City. In one of those children's crusades strode a thirteen-year-old handloom weaver named Alexander Bain. Forced to abandon his formal education two years earlier, he was spending his scant hours away from the loom teaching himself higher mathematics from the Books of Euclid. In a few more years, his drive and brilliance would begin to propel him out of poverty. At the age of thirty-seven he would publish a book that would place him among the most influential philosophers of the Victorian age.

Bain became famous for turning psychology into a science. In a series of groundbreaking textbooks, he showed how thought might arise from neural physiology, and opened up the human brain and mind to laboratory study. He was also a devout utilitarian.[1] Jeremy Bentham was his lifelong hero and John Stuart Mill his most cherished friend. From his perspective, utilitarianism was self-evidently the ethics of physical reality. In a pair of weighty volumes published in the 1850s, he brought pleasure-pain psychology, sensory-motor physiology, and utilitarian morality together into one seamless whole. Bain was the most important link between Bentham and the American behaviorists, providing the basic theoretical framework for all pleasure-pain psychology up to the 1970s.

Bain was born in 1818 into harsh Scottish poverty. His father was a handloom weaver who fought as a loyal British patriot in the Napoleonic Wars. After his discharge from the army, Bain senior married a young woman and fathered eight children in rapid succession. Handloom weaving was a good trade when the couple married, but the mechanization of the textile industry meant that the money for piecework steadily fell. Eventually he found himself working thirteen to fifteen hours a day, trying to provide for a growing family as the Industrial Revolution robbed him of his livelihood. Three of his eight children died in infancy; only Alexander, the second child, saw the age of forty.[1]

Alexander did more than just survive. He blossomed wildly, eventually commuting a threadbare education into a meteoric academic career. His first teacher was an "old dame, who taught only reading." After that he took lessons with a series of college students in tiny local schools. When he was seven years old, one of these young men told his parents that he was unusually gifted.[2] But no amount of surprising intellectual ability was going to fill the family coffers, and Bain's formal education ended where for most modern British children it begins in earnest. At age eleven, he left school to earn a living. His first job was as an errand boy to an auctioneer, assisting at sales and opening and sweeping the rooms in the mornings. After two years he found the responsibility too much and left. "My father now put me to the loom, where I continued for the next five years, which was also the period of my self-education."[3] Aged thirteen he became a political activist and learned that Bentham, advocate of universal suffrage, was a hero of the workingman.

Bain's account of his five years at the loom, between the ages of thirteen and eighteen, is a testament to the raw hunger of his intellect. After a long day weaving, he would turn his mind to mathematical problems, working his way through the Books of Euclid at the prodigious rate of one every couple of days. For companionship he turned to the two sons of a local blacksmith, "whose smithy was an agreeable lounge, especially in winter evenings." The father had amassed a library of metaphysical and scientific books, and the sons, self-taught mathematicians like Bain, had built their own telescope, through which the three boys studied the satellites of Jupiter and the craters of the moon. Bain debated free will and idealism in the smithy salon, always coming down on the side of materialism and determinism.[4] Other intellectual mentors at the time included an "eccentric genius" watchmaker, who allowed Bain to look at his precious copy of Newton's *Principia*, and a mathematics teacher, a "singular specimen of humanity . . . less than four feet high, but erect in person, and of a commanding, not to say fierce, expression," to whom he paid a fee for evening lessons.[5]

Overhearing Bain in conversation in a bookseller's shop one day, a local minister was impressed by the intellectual precocity of the seventeen-year-old weaver and suggested that he prepare himself for university entrance, offering to tutor him in Latin. After a year of lessons, Bain took the examination and was offered places in two of Aberdeen's ancient colleges (both of which eventually became part of the University of Aberdeen). His academic career had begun, and the hungry-minded young man never looked back, vying with his more privileged classmates for top places in the prize lists. At this juncture, he embarked on his lifelong study of the operations of his own mind, which soon became "incessant and overmastering."[6] Attempting to catch his powerful brain in the act of thinking, he searched for lawlike regularities in his trains of thought and association.

In 1840 Bain began writing for the *Westminster Review*. He was now a college graduate of twenty-two, lecturing for the Mechanics Institute and tutoring the offspring of the local aristocracy. Egged on by a well-connected friend, he submitted an article on two new inventions, the electrotype (a method for making duplicate plates for printing) and the daguerreotype (an early form of photography). Success emboldened him to follow with another entitled "Accurate Investigation on the Properties of Matter." Already a devout utilitarian, Bain began to ready himself to meet John Stuart Mill. After brushing up on all the back issues of the *Westminster Review* and making himself thoroughly familiar with Mill's thought, he embarked on an adulatory correspondence with the Colossus of Liberalism.

In April 1842 Bain made his momentous first trip to London to pay tribute to his idol. The day after he arrived, he walked to the headquarters of the East India Company, a "dingy, capacious, and venerable building in Leadenhall Street" where Mill labored at the administration of the empire. The first sight of Mill, standing at his desk at the end of the long, light-filled room, was clearly emotionally charged for the twenty-four-year-old autodidact and later stirred him to one of his rare flights of description:

> I am not likely to forget the impression made upon me, as he stood by his desk, with his face turned to the door as we entered. His tall slim figure, his youthful face and bald head, fair hair and ruddy complexion, and the twitching of his eyebrow when he spoke, first arrested the attention: then the vivacity of his manner, his thin voice approaching to sharpness, but with nothing shrill or painful about it, his comely features and sweet expression—would have all remained in my memory though I had never seen him again.[7]

For the next five months, Bain met Mill after work at India House twice a week, and walked with him back to his home in Kensington Square, discussing all of Mill's projects and schemes. He also dined with him on occasion and was introduced to the leading members of the *Westminster Review* circle. There were rumblings of a job teaching philosophy in Bombay, but nothing came of it.

In the winter following this star-studded interlude, Bain sealed his status as Mill's protégé by writing a review of the great man's *Logic* for the *Westminster Review*. Thereafter, he spent his summers in London, building up a network of contacts in the liberal literary world. During the dark months of the year he pursued a career as a philosophy lecturer in Aberdeen, cranking out sundry articles for London journals on scientific and philosophical topics. As his immersion in utilitarian circles deepened, Bain began to work on his magnum opus, a scientific psychology that would bring pleasure-pain psychology and sensory-motor physiology together in a materialist account of the human condition. *The Senses and the Intellect* was published in 1855; its companion volume, *The Emotions and the Will*, appeared three years later. These books, composed when he was in his thirties, were Bain's monuments. To keep up with developments in physiology, he continued to make substantial revisions to them all his life. Successive editions appeared at regular intervals until the end of the century, serving as the standard British psychology textbooks for almost fifty years.

The Senses and the Intellect opened with a fifty-page exposition of neuroanatomy and sensory-motor physiology. Consisting of extracts from various standard anatomy textbooks, it was far from original, but it was an important declaration of his commitment to a material foundation for human psychology. After the anatomy and physiology, Bain plunged into a section on "the Senses": "all that is primitive or instinctive in the susceptibilities and impulses of the mental organization."[8] "Intellect," "Emotion," and "Will," the other three primary elements of his scheme, he assembled upon this physiological foundation. This bottom-up organization—starting with anatomy and physiology, moving on to the primitive and instinctive senses, and only then discussing higher psychology—was a radical statement of biological materialism. By asserting that thought had a physical substrate, the books opened up the human mind to laboratory investigation and secured Bain's status as the father of British scientific psychology.

Above all, Bain's psychology was based on the primal experiences of pleasure and pain. "In joyful moods," he observed, "the features are dilated; the voice is full and strong; the gesticulation is abundant; the very thoughts are richer. In the gambols of the young, we see to advantage the coupling of the two facts—mental delight and bodily energy." After painting this pretty scene, he asked the reader to imagine its abrupt reversal: "Introduce some acute misery into the mind at that moment, and all is collapse, as if one had struck a blow at the heart." Although "blow at the heart" was hardly scientific, he concluded the passage with an appeal to medical science: "A medical diagnosis would show, beyond question, that the heart and lungs were lowered in their actions just then; and there would be good grounds for inferring an enfeebled condition of the digestive organs."[9] Pleasure-pain psychology involved the whole body and brain, working together as a single mechanism, producing thought, feeling, and behavior as primal reactions to delight and misery.

In accordance with his utilitarian psychology, Bain endorsed

Benthamite ethics. For him, there was simply no viable alternative. Utilitarianism trumped every other philosophy because it set up "an outward standard in the room of an inward, being the substitution of a regard to consequences for a mere unreasoning sentiment or feeling."[10] Bain's textbooks appeared right after the publication of Dickens's *Hard Times*, that anti-utilitarian fable in which Sissy Jupe, a young, naïve, golden-hearted circus performer, stands in for everything warm, human, and virtuous that was missing from the philosophies of Bentham and Malthus. Bain ended his section on the emotions with a dig at the Sissy Jupes of this world, accusing them of "a sentimental preference, amounting to the abnegation of reason in human life."[11] With these books, he finished the work that Bentham and James Mill had started. Pleasure-pain psychology and utilitarian ethics were now squarely grounded in anatomy and physiology. There were no chinks in the smooth walls of the utilitarian edifice where immaterial mind might seep in. Bentham's sketch of a discipline of "moral physiology" had been completely filled in.

THE LAW OF ADAPTATION

For all its rigorous basis in biology, however, Bain's theory of mind was lacking one important ingredient. He offered no account of how his utilitarian self might have come into being. Humans were clearly part of the brute creation, but what, exactly, was their family relationship to the other animals? An 1859 review of one of his books by the writer and philosopher Herbert Spencer answered the question. One by one, Spencer took Bain's categories and recast them in evolutionary terms. This move gave tremendous explanatory backbone to pleasure-pain psychology, accounting for appetites and aversions as "adaptive" physical mechanisms. According to Spencer, every sentient being—from mollusk to member of Parliament—had evolved a system of pain and pleasure to drive it away from harmful things and draw it toward helpful ones. Spencer's evolutionism completed the list of ingredients baked hard into twentieth-century behaviorism.

There were many parallels between these two giants of Victorian psychology: Bain and Spencer were born within two years of one another and died the same year; they mixed in the same circles of reform-minded intellectuals centered on the *Westminster Review*; they shared the conviction that they had ruined their health by overtaxing their brains. Their personal friendship was slow to ripen, but they engaged in a long and respectful dialogue in their published works. In 1904, the year after both men died, their respective autobiographies appeared almost simultaneously. By the end of the nineteenth century, their names were yoked together as codevelopers of a "law of pain and pleasure," showing how evolution, behavior, and learning operated under the sway of Bentham's two sovereign masters.

Spencer was born in 1820, just two years later than Bain, in the heart of the Industrial Enlightenment. His schoolteacher father was secretary to the Derby Philosophical Society, younger cousin of the better-known Lunar Society, a discussion club featuring lectures on phrenology, chemistry, electricity, and pneumatics. Raised in the optimistic Enlightenment tradition that had spread outward from Joseph Priestley and the Lunar Men to dozens of doctors, naturalists, and teachers like his father, Spencer never lost his faith that science and technology were hustling humankind down the path to perfect happiness.

After being homeschooled by his father and uncles, Spencer became a civil engineer, working on various problems in the burgeoning railway network. At twenty-one he quit his job in the hope of taking out a patent on an electromagnetic engine, the failure of which left him living at home for three years, dabbling in suffrage politics, writing for *The Nonconformist*, and pursuing various mathematical, phrenological, engineering, and natural historical questions in a more or less desultory fashion. Eventually he realized that the uncertainties of the inventor's life were not for him and began to cast around for other work. A letter of introduction to the editor of the *Economist* resulted in an invitation to take up the subeditorship, a position with modest pay but some alluring fringe benefits, including accommodations in a small apartment attached to the office on the Strand and free opera tickets.

As part of his wide-ranging if unsystematic self-education, Spencer had already read some Bentham, writing to a friend that he intended "to attack his principles shortly."[12] During his tenure at the *Economist* he launched the attack, in the form of a five-hundred-page anti-utilitarian treatise called *Social Statics; or, The Conditions Essential to Human Happiness Specified and the First of Them Developed.* The book opened with a chapter assaulting the utilitarian pursuit of the greatest happiness for the greatest number. First, Spencer argued, we cannot know what makes men happy, as tastes are too various. Second, we cannot know what the consequences of our actions will be, as causation is too complex. To illustrate the second difficulty, he ran gleefully down a list of the un-intended consequences of government meddling: a Bavarian prohibition on men marrying before they could support a family only led to a rapid increase in illegitimate births; forty warships sent to the west coast of Af-rica to disrupt the slave trade only provoked an intensification of its cru-elty; and so on. His favorite example of utilitarian futility and failure was the Benthamite/Malthusian Poor Law reforms: "For the production of the 'greatest happiness,'" he announced sarcastically, "the English people have sanctioned upwards of one hundred acts in Parliament."[13]

Against this futile forecasting of consequences, Spencer argued that there were, in fact, a few simple rules of right conduct. These arose as a matter of logical necessity from our condition as social beings.[14] He listed four: "justice," "negative beneficence," "positive beneficence," and "self-care."[15] Note the uncanny resemblance between Spencer's four rules and the four principles of contemporary biomedical ethics: autonomy, non-maleficence, beneficence, and justice. Although justice is the first principle in Spencer's order and the last in the modern ranking, the order of priority was not, in fact, reversed. For Spencer, "justice" was "the gen-eral proposition, that every man may claim the fullest liberty to exercise his faculties compatible with the possession of like liberty by every other man."[16] His first principle, in other words, was much closer to the modern notion of autonomy than to the distributive concerns of the fourth (and mostly neglected) modern principle of "justice." As for Spencer's principle of "self-care," the one that does not appear in the modern version, we will return to that in the afterword.

For Spencer, the only legitimate role of the state was to support, adju-dicate, and maintain individual freedom. His libertarianism arose from a devout belief in moral, scientific, and biological progress. Inflated with his Victorian self-confidence, the cheerful progressivism of his Enlight-enment forebears had expanded into an extreme faith in rapid and inex-orable improvement of every kind. As he pondered the sweeping changes taking place around him, he became convinced that all of life was under-going speedy evolution toward physical and spiritual perfection.

In accordance with his belief in progress, Spencer proposed a biological "law of adaptation," according to which all organisms, in the space of a few generations, adapted to surrounding conditions: "If exposed to the rigour of northern winters, animals of the temperate zone get thicker coats and become white. . . . Cattle which in their wild state gave milk but for short periods, now give it almost continuously." Since mankind was similarly evolving, the existence of any form of government was only a temporary state of affairs, destined for moral obsolescence within a few generations: "As civilization advances, does government decay. . . . What a cage is to the wild beast, law is to the selfish man."[17]

The following year, 1852, Spencer weighed in on the Malthusian debate, producing an article for the *Westminster Review* outlining his own "Theory of Population." As usual, he saw the issue in terms of biological adaptation, arguing that fertility declined with the development of the "nervous centers." As the human species became more civilized and intelligent, he urged, the fewer children would be born, until "on the average, each pair brings to maturity but two children." At that point, the relationship between man and his environment would have reached an optimal equilibrium.

In the course of expounding this vision, Spencer threw out a few remarks on population and competition, on the basis of which he later claimed to have discovered the principle of natural selection seven years before the publication of Darwin's *Origin of Species*: "As those prematurely carried off must, in the average of cases, be those in whom the power of self-preservation is the least, it unavoidably follows, that those left behind to continue the race are those in whom the power of self-preservation is the greatest."[18] But Spencer's evolutionism was a far cry from the slow, directionless, Malthusian war of all against all proposed by his rival. It remained incorrigibly Lamarckian and celebratory, a secular update of Joseph Priestley's spiritual belief in the rapid upward climb of all beings toward a state of perfect enjoyment.

In 1855 Spencer published *The Principles of Psychology*, in which he traced the evolutionary steps by which the human mind developed from the "primordial irritability" of the simplest organisms. He sketched how the habits, instincts, and emotions of every animal species grew out of the repetition of pains and pleasures, which consolidated themselves into innate responses. The process was ongoing and complex. Even human political instincts could be understood in evolutionary terms, including the love of liberty and "the sentiment of personal rights."[19] For Spencer, natural rights were no more or less than biological instincts, the product of the evolution of a gratification-seeking social primate, moving inexorably toward the ever-greater adaptation of means to ends.

In an 1860 review of Bain's *The Emotions and the Will*, Spencer opined that while the author had done a good job of describing human psychology in materialistic terms, he had left out a factor of paramount importance: "Nothing like a true conception of the emotions is to be obtained until we understand how they are evolved."[20] Nothing about pain and pleasure or habit and instinct made any sense, Spencer insisted, unless it was understood in the light of evolution. In the next edition of his book, Bain fought back. He acknowledged that human psychology was a product of evolution but scoffed at Spencer's belief in "the hereditary transmission of acquired mental powers and peculiarities."[21] By this time, Darwin's *Origin of Species* had been published, and Bain decided in favor of the blind, brutal, and slow action of natural selection rather than Spencer's rapid, progressive law of adaptation.

Posterity has been kinder to Bain's Darwinism than Spencer's Lamarckism, but they both relied on their own most intimate experience to decide the matter. Spencer's middle-class upbringing had convinced him that every quirk and prickle of his personality was inherited directly from his argumentative aunts and uncles.[22] Bain's experience of the intellectual gifts that lifted him from humble origins to dizzying heights made heredity a different sort of puzzle for him: "Men of genius in all countries spring indiscriminately from every order of the community," he pointed out. "Shakespeare and Cromwell, Burns and Scott, rose at one bound from the humblest mediocrity; and their genius was uninherited." Against the idea of inherited intelligence, Bain cited a study suggesting that the best chances of mental vigor lay "with the second child of a mother marrying about twenty-two," exactly describing the circumstances of his own birth.[23]

On January 9, 1878, worried that his chronic ill health was soon to render him incapable of working, Spencer summoned his amanuensis to his bedside and began to dictate the great work of evolutionary ethics that he had spent a lifetime pondering. In fewer than three hundred tightly argued pages, he laid out the reasons that pleasure-pain psychology formed the only scientific basis for moral reasoning and explained why utilitarianism had failed to make good on its Epicurean premise. Calculating the misery or happiness produced by every action is impractical, unsustainable, and unscientific, Spencer argued. Only by deducing rules of right conduct from an evolutionary understanding of health, pleasure, and happiness, and then sticking to those rules, will the human race learn to live with itself in harmony. The main message of the book was that goodness was always pleasurable, and Spencer made a brisk survey of ethical systems, showing how each could be nudged toward this conclusion with a dose of clarity mixed with some well-placed sarcasm.[24]

By the time he wrote *The Data of Ethics*, however, Spencer's optimism was looking a bit outmoded. Early on in his career, he had issued a challenge to any reader inclined to be skeptical about the imminent triumph of liberty. All dreary pessimists should "visit the Hanwell Asylum for the insane" and "witness to their confusion how a thousand lunatics can be managed without the use of force."[25] "Moral treatment"—the idea that an institutional environment of kindness, reason, and orderliness could cure insanity—was a potent symbol of Enlightenment progress, and its early successes seemed to herald a new dawn of human improvement.[26] But in the 1860s the asylums began to fill up with the casualties of industrialization, and the nonrestraint method found itself hopelessly unequal to what seemed like an epidemic of mental illness. The Enlightenment optimism that had fueled moral treatment crumbled, giving way to genetic fatalism and fears of degeneration.

An irony of this collapse of Spencerian optimism was that it gave Spencer's theories a mighty scientific boost. In the 1870s the rising tide of mental illness among the poor of industrial Yorkshire provoked the young director of the West Riding Pauper Lunatic Asylum to turn to basic science for answers. In 1873 he brought in a promising young experimenter called David Ferrier and set him up in a purpose-built laboratory on the asylum grounds. There, Ferrier opened up the skulls of dozens of animals and birds, and probed their brains with electrodes, to map the sensory motor system on the cortex. Ferrier was a disciple of Bain and Spencer, and he used their work to explain how sensory information turned into motor action via the pleasure-pain system. By bringing Bain and Spencer's theories into the laboratory, Ferrier turned their speculative psychology into an experimental science, driving utilitarianism deep into the emergent discipline of neurology.

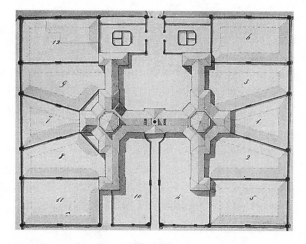

Figure 9. West Riding Pauper Lunatic Asylum, architect's plan and main building façade. Inspired by Bentham's Panopticon, the asylum's main building wraps the industrial architecture of penal reform in a skin of Georgian elegance. The two octagonal towers house staircases that provide a panoptic view of the whole interior.

THE WEST RIDING PAUPER LUNATIC ASYLUM

Just outside the town of Wakefield in Yorkshire, in the midst of a warren of prefabricated National Health Service offices, stands the original building of the West Riding Pauper Lunatic Asylum (saved by a conservation order and now turned into apartments). Construction began in 1815, only six months after the Battle of Waterloo, and much of the labor was supplied by POWs from Napoleon's armies. The façade of pale gray brick, built in a neoclassical style, with shallow recesses housing large, white-painted sash windows, is simple to the point of severity, evoking the quiet piety of Samuel Tuke, the Quaker madhouse reformer who was the moving spirit behind the asylum.[27]

In keeping with Tuke's spiritual purpose, the tall, unbarred windows give the "face" of the building an expression of luminous frankness. In his instructions to the architects, Tuke remarked that "the best general security against injury to windows is to make them easy of access . . . to break windows when they are completely within reach is to achieve nothing."[28] This genial observation exemplifies his reformist philosophy. He had inherited his views from his grandfather, the Quaker tea merchant William Tuke, who founded the Retreat at York, the first asylum in England to espouse the healing effects of kindness. A gay little cupola above the central pediment of West Riding (hiding an exhaust chimney) seems to embody the Tukes' conviction that a regime of rational conversation, fresh air, and gentle occupation could restore the insane to reason.

Behind the lucid simplicity of the façade, however, stand two wide, flat-topped octagonal towers. These are the surveillance posts for the asylum. By 1815 Tuke's Quaker faith in moral treatment had been supplemented by Bentham's philosophy of the controlling gaze: "The regulations of an Asylum should establish a species of *espionage*,"[29] he wrote, marking him as a member of the generation of reformers who labored in Bentham's long shadow. Accordingly, the octagonal towers contained winding staircases with balconies from which the whole operation could be surveyed: "From a landing about halfway between this and the next story all that go up or down have a complete view of what is going on."[30] Even now, the towers seem at odds with the neoclassical façade. The surveillance posts concern the management of the human animal; the unbarred windows are for the treatment of the human soul.[31]

In his "Panopticon Letters" Bentham breezily recommended that the inspection principle be applied to all new madhouses, where it "would render the use of chains and other modes of corporal sufferance as unnecessary in this case as any."[32] Samuel Tuke knew better: assuming lunatics to be beyond the reach of such rational management, the principal target of surveillance at West Riding was not the inmates but the employees. As a member of the third generation of madhouse reformers in his family, he was well aware of the stresses and temptations attendant on working in a pauper lunatic asylum. In his scheme "inspection must form a check if not a preventative to improper conduct in the servants. The certainty that their conduct is under the inspection of everyone who chooses to look at them as they pass and repass must necessarily have a beneficial tendency."[33] Tuke's philosophy of surveillance was moderate in all respects, and he explicitly distanced it from Bentham's design, musing that it was possible "to carry our views of easy inspection too far; and I think it would be doing so to adopt some of the panopticon plans, in which a centre room, lighted from above, and enclosed on all sides, by the apartments of patients, is appropriated to the master and mistress of the family."[34]

For about a decade after its founding, West Riding's regime of moral treatment plus diluted panopticism seemed to work. The first director was William Ellis, a devout Methodist who brought Christian principles to bear on every aspect of asylum life. In his *Treatise on the Nature, Symptoms, Causes and Treatment of Insanity*, he defines the "most essential ingredient" of moral treatment as "constant, never-tiring, watchful kindness."[35] Ellis pioneered the use of what would now be called occupational therapy, putting the patients to work growing food and weaving cloth, combining therapeutic and economic advantages with Benthamite frugality. The asylum was a total institution, complete with its own brewery, shoemaker, printing press, laundry, and farm. There was a gasworks on the grounds, converted to the manufacture of coal gas when whale oil began to run out as a result of overfishing by the Whitby whaling fleet. The whole complex was designed as a microcosm of the wider world, purged of the disorder and cruelty that had created mental derangements, a utopian Christian community of gentleness and spiritual uplift.

In 1829 William and Mildred Ellis's community began to grow beyond the bounds of the original building, and the first of many extensions was undertaken. In 1831 the Ellises left for a post at a grander asylum and passed the baton to Charles Corsellis, who had to undertake a massive expansion of patient accommodation as industrialization, urbanization, and epidemic disease drove more and more people to mental breakdown. In 1837 Corsellis extended the work program, causing vast weaving sheds to be built on the asylum's grounds, boasting in his annual report that the "beneficial results . . . may be seen in the tears of gratitude shed by hundreds of restored sufferers."[36] Under the Ellises, the asylum had been a utopian community modeled on the bourgeois family; under Corsellis, it became a microcosm of the industrial economy.

Even as inmate numbers grew, Corsellis tried to uphold the spirit of moral treatment. He was the last director to appoint his wife matron and to live in the asylum as a benevolent patriarch. Here is his description of the basic structure of institutional life, penned in 1838 in his even copperplate hand, worth quoting in full for its evocation of the simple Quaker rhythms of the day:

> The Patients are all paupers, their respective parishes paying for each 6s a week. This sum defrays every expense. They are fed, lodged, and clad alike, wearing a dress of grey woollen cloth, which is woven and made up by themselves; they rise at six a.m. in the summer, and seven in the winter, and all those who are in a fit state (of whom there are a great number) attend with such servants as can be spared at morning prayers, precisely at eight o'clock. They breakfast on milk pottage and bread at half past eight. At nine o'clock, the gardeners, farmers, laundry women, etc. select those who by previous arrangement with the Director have been fixed on, for their several occupations, and commence work. At eleven the workers have a luncheon of bread and three-quarters of a pint of table beer. They dine at one. Their dinners are one day, meat, yeast dumplings and potatoes, and half-a-pint of beer; the next, soup with potatoes and dumplings, alternately. At two, work is resumed, and at four, a luncheon is distributed, similar to that in the forenoon. At seven they have supper of milk pottage and bread. At eight, the bedroom doors and window shutters are carefully locked, the clothes folded and placed on the outside of each door.[37]

In 1839 he proudly noted that "of the 201 male patients now in the house, only one is under restraint." He reserved some scorn for the Benthamite naiveté of his predecessor, noting that "the result of many years attention to this subject, has led to the conviction that a mild and judicious restraint can never be supplied by any 'surveillance.'"[38]

By the time Corsellis wrote these words, however, ruthless industrialization was creating problems in the face of which moral treatment seemed helpless. Even as he congratulated himself on the success of occupational therapy, he lamented the economic pressures and destructive temptations grinding down the British working class: "the masses congregated in the houses, yards, and factories of our manufacturing towns" and "the calamitous effects of the everywhere to be found 'Beer Houses.'"[39] Faced with an overwhelming influx of new inmates, he did what he could to improve the treatment regime. In 1838 he noted that a patient had died after being bled, and for a few years leeches quietly disappeared from the expense accounts.[40] In 1839 he drew up a table of the causes of insanity for all patients, arranged in alphabetical order from "Anxiety of mind," "Asthma," and "Avaricious disposition" to "Studying Astrology," "Unkindness of husband," and "Venereal disease."[41] In 1840, taking advantage of the Benthamite Anatomy Act, he constructed a list "Showing the Average Weight of Brain in the cases of 90 Male and 46 Female Patients, who have died in the Asylum."[42]

Nothing proved equal to the rising tide of lunacy. By midcentury a profound therapeutic pessimism began to afflict madhouse reformers all across the nation. Moral treatment may have turned asylums into relatively clean and orderly places, but it did nothing to reduce the number of admissions. Instead, inmate numbers grew catastrophically, staff turnover quickened, and the nonrestraint system began grimly to revert to the same custodial function that it was supposed to have rendered obsolete. As the industrial revolution gobbled up ever-vaster tracts of Yorkshire, the Arcadian optimism that had animated West Riding's founding broke down entirely. In 1866 the directorship passed to James Crichton-Browne, who responded to its overwhelming problems by transforming West Riding into an internationally celebrated center for neurological research.

Like Samuel Tuke, Crichton-Browne was born into the madhouse trade. His father was an asylum reformer, and he was brought up in a house on the grounds of a progressive asylum in Scotland, right up until he left for medical school at the University of Edinburgh. After a series of brief asylum appointments, he landed the top job at West Riding at the tender age of twenty-six. The problems he encountered were huge. By March 1868, for example, 1,229 patients were in residence. The asylum had originally been designed to house 150. At a meeting in 1869 he complained that "there is something exceptional in the enormous apparent increase of lunacy which is now taking place in Riding."[43] A few years later he lamented being "submerged" by a "wave of lunacy."[44] But he was young and full of confidence, and he plunged headfirst into the wave, passionately asserting that he could never "be content with a system which would simply provide convenient storage for heaps of social debris."[45]

Crichton-Browne's attack was two-pronged. Convinced of the therapeutic effects of pleasure, he directed the construction of large Turkish baths for the patients, as well as the installation of a theater, where facetious pantomimes and melodramas were staged. These performances used a combination of professional, amateur, and inmate actors, including, astonishingly, one show in which W. S. Gilbert of Gilbert and Sullivan fame appeared, with sets painted by the famous neoclassical artist Alma Tadema.[46]

The other prong of Crichton-Browne's approach was a turn to basic science. To that end, as soon as he arrived he retained the services of a pair of "Clinical Clerks" to draw up case reports. He began to take photographs of the inmates, some of which found their way into Charles Darwin's book *The Expression of the Emotions in Man and Animals*. Like Corsellis, he made sure to examine the brains of patients who died in the asylum, hoping to find visible lesions or abnormalities in their neural tissue that would explain their afflictions.[47] Beginning in 1871 he also convened an annual meeting of leading pathologists, clinicians, psychologists, and physiologists to discuss the application of medical science to problems of the nervous system. Their papers were collected annually in the *West Riding Lunatic Asylum Medical Reports*. The first article, entitled "Cranial Injuries and Mental Disease," by Crichton-Browne himself, was based on the autopsies and case histories that were beginning to be systematized under his leadership.[48]

In 1872 Crichton-Browne convinced the trustees of the asylum that he needed a full-time pathologist on staff, an innovation he described as "a somewhat momentous step in the march of scientific progress in the Lunatic Asylums of this country." Then he unrolled his plans for the building: "To give full scope to the energies of such an Officer as a Pathologist, and to utilize to the highest advantage his labours, a Pathological Institute or detached building, containing a museum, laboratory, microscopic, photographic, and lecture rooms, is certainly requisite."[49] Shortly thereafter, a small building between the laundry and the gasworks was erected to house a morgue, an autopsy room, and a physiology laboratory.

Mysteriously, after the first pitch that Crichton-Browne made for the laboratory, the project disappears from the institution's archive. The building was certainly constructed: it appears on the huge ground plan of the asylum that was drawn up in 1888. However, no mention of the expenses associated with it, including the purchase of experimental animals, appears in the accounts for any year. Moreover, the Committee of Visitors reports for 1874 and 1875 are missing completely. As we will see later in the chapter, David Ferrier, the physiologist who worked there, was involved in a high-profile animal experimentation scandal, which may account for the absence of any trace of his laboratory from the records of the asylum.[50] The complete lack of a paper trail gives the place a certain ghostly quality, but it was there that Ferrier turned utilitarianism into a science of animal behavior.

DAVID FERRIER

Born in Aberdeen in 1843, David Ferrier studied classics and philosophy under Alexander Bain at the University of Aberdeen. It was Bain who introduced the young man to experimental psychology and who sent him to Heidelberg to study under the German pioneers of "psychophysics," a forerunner of cognitive science. After one year abroad Ferrier came back and completed his medical training in Edinburgh. In 1870 he moved to London and began work as a consultant at the National Hospital for Paralysis and Epilepsy in Queen Square, the first institution to specialize in seizure disorders.[51]

The year Ferrier arrived at the hospital, the leading clinician, John Hughlings Jackson, a giant of nineteenth-century British neurology, had just made his great theoretical breakthrough. He had realized, on the basis of clinical observation alone, that the "march" of a partial seizure—the way that it would often start in one place, such as the thumb and forefinger, and then progress in an orderly fashion along adjacent parts of the body—was a clue to the way the brain was organized. The very regularity of the progress of a seizure suggested that the human body was "represented" in the cerebral cortex in some invariant fashion. The order in which body parts succumbed to epileptic movement must correspond to the way in which they were represented in the brain. Larger areas of the cortex were devoted to the most functionally complex parts of the body—lips, tongue, hands—which were thus more susceptible to the abnormal electrical discharge causing seizures.[52]

In 1870 two Prussian experimenters published the results of a series of experiments that seemed to confirm these conjectures about the layout of the cortex. Using an apparatus devised by one of them for administering therapeutic electricity to patients, the pair conducted a series of experiments on stray dogs, opening their skulls and electrically stimulating the surface of their brains to see if they could produce bodily movements.[53] With these results they were able to produce the first "map" of the motor area of any species' brain.

British scientists reacted to the Prussian experiments with a mixture of excitement and panic—excitement that the brain seemed to be yielding its secrets, and panic that the Germans might take credit for being the first to probe the orderly localization of motor function. In 1873 Jackson, Ferrier, and the other rising stars in British neurology gathered for the West Riding Lunatic Asylum's annual meeting. The discussion turned to the Prussian work on dogs, and Crichton-Browne invited Ferrier to come to the asylum to conduct a series of experiments in the newly built laboratory, replicating the Germans' results on a much larger scale.[54]

The project was completed in record time. Using animals provided by Crichton-Browne, Ferrier conducted a series of electrical stimulations on the brains of rabbits, dogs, cats, guinea pigs, and birds. In 1873 the *West Riding Lunatic Asylum Medical Reports* contained an article summarizing his conclusions, arguing that stimulation of the cortex produced consistent and predictable movements in every animal of the same species.[55] This set the stage for the next phase of Ferrier's program: the extension of his experiments to primates. In the spring of 1873, with a grant from the Royal Society, he undertook a series of experiments electrically stimulating the brains of thirteen macaque monkeys. Using primates allowed him to suggest "that comparison may be made with . . . the human brain."[56] This was a radical statement: Ferrier was asserting a close family relationship between humans and monkeys.

Transatlantic support arrived promptly when an American physician, Roberts Bartholow, proved Ferrier's point about human-animal continuity with a series of electrical-stimulation experiments on a human subject. In his report, Bartholow described how a thirty-year-old Irish housemaid named Mary Rafferty—"rather feeble-minded . . . cheerful in manner"—had presented herself to the Good Samaritan Hospital in

Cincinnati with an ulcer on her scalp so severe that it exposed part of her cerebral cortex. Inspired by Ferrier's animal studies, Bartholow resolved to exploit this opportunity. His plan was to "conduct tentative experiments . . . on different parts of the brain, proceeding cautiously."[57] "Unfortunately," he recounted, "owing to a rapid extension of disease to the left hemisphere, these details were not fully carried out."

This explanation was disingenuous to say the least. Here is Bartholow's own description of the "extension" of Mary Rafferty's "disease":

> When the needle entered the brain substance, she complained of acute pain in the neck. In order to develop more decided reactions, the strength of the current was increased by drawing out the wooden cylinder one inch. When communication was made with the needles her countenance exhibited great distress, and she began to cry. Very soon the left hand was extended as if in the act of taking hold of some object in front of her; the arm presently was agitated with clonic spasms; her eyes became fixed with pupils widely dilated; lips were blue and she frothed at the mouth; her breathing became stertorous, she lost consciousness and was violently convulsed on the left side. The convulsion lasted five minutes and was succeeded by coma.[58]

Rafferty recovered sufficiently to be subjected to another two rounds of stimulation over the course of the next three days, rendering her "stupid and incoherent." On day four of her ordeal, another five-minute seizure was succeeded by "profound unconsciousness." The next section of the article details the results of her autopsy.[59] For Bartholow, these experiments proved Ferrier's claim about the identity of monkey and man. Mary Rafferty—feeble-minded, poor, and Irish—had become an experimental primate.

Rafferty's death became a test case for the ethics of human and animal experimentation. The American Medical Association publicly censured Bartholow for the work. The full text of the AMA resolution was published in *Cincinnati Medical News* under the headline "Human Vivisection." In Britain the medical establishment was more receptive. Ferrier was the first to publish a report of the experiments. "Whatever opinion may be entertained as to their propriety," he said, "they furnish facts of great interest in relation to the physiology of the brain."[60] Ten days later the *British Medical Journal* summarized the research without any comment as to the rights and wrongs of the case.[61]

Later in the year Ferrier gave another talk to the Royal Society about his macaque monkeys. "Lively, active and intelligent," drinking sweet tea and eating bread and butter in front of the fire after their surgeries, grunting with recognition when called by name, curious, impatient, and sociable, his monkey research subjects seemed at least as human as the "cheerful, feeble-minded" housemaid who underwent a similar ordeal in Cincinnati.[62]

In 1876 Ferrier published his magnum opus, *The Functions of the Brain*. The book became an instant classic and has been celebrated by generations of neuroscientists as establishing cerebral localization on a scientific footing for the first time. For our purposes, what is interesting about *The Functions of the Brain* is the way it translated Bain and Spencer's pleasure-pain psychology into the language of experimental science. Where Bain and Spencer introspected, Ferrier vivisected, rebuilding their psychological theories on a foundation of animal pain and paralysis.

In chapter 11—"The cerebral hemispheres considered psychologically"—Ferrier announced that his "psychological principles . . . coincide with those expounded by Bain and Herbert Spencer."[63] The chapter opened with a half-page quotation from Bain about the materiality of mind: "We have every reason to believe that there is, in company with all our mental processes, *an unbroken material succession*."[64] Ferrier then went on to ground his theory of sensation and memory in a pleasure-pain account of learning:

A hungry dog is impelled by the sight of food to seize and eat. Should the present gratification bring with it as a consequence the severe pain of a whipping, when certain articles of food have been seized, an association is formed between eating certain food and severe bodily pain; so that on a future occasion the memory of pain arises simultaneously with the desire to gratify hunger, and, in proportion to the vividness of the memory of pain, the impulse of an appetite is neutralised and counteracted. The dog is said to have learned to control its appetite.[65]

With this description of a dog "learn[ing] to control its appetite," Ferrier laid out the basic principles of behaviorist psychology. All behavior was

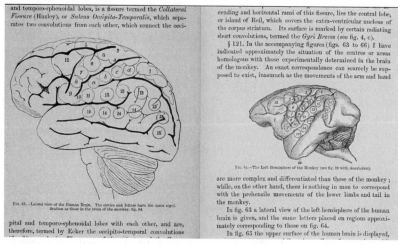

and temporo-sphenoidal lobes, is a fissure termed the *Collateral Fissure* (Huxley), or *Sulcus Occipito-Temporalis*, which separates two convolutions from each other, which connect the occi-

cending and horizontal rami of this fissure, lies the central lobe, or island of Reil, which covers the extra-ventricular nucleus of the corpus striatum. Its surface is marked by certain radiating short convolutions, termed the *Gyri Breves* (see fig. 4, c).

§ 121. In the accompanying figures (figs. 63 to 66) I have indicated approximately the situation of the centres or areas homologous with those experimentally determined in the brain of the monkey. An exact correspondence can scarcely be supposed to exist, inasmuch as the movements of the arm and hand

FIG. 64.—The Left Hemisphere of the Monkey (see fig. 20 with description).

are more complex and differentiated than those of the monkey; while, on the other hand, there is nothing in man to correspond with the prehensile movements of the lower limbs and tail in the monkey.

In fig. 63 a lateral view of the left hemisphere of the human brain is given, and the same letters placed on regions approximately corresponding to those on fig. 64.

In fig. 65 the upper surface of the human brain is displayed,

FIG. 63.—Lateral view of the Human Brain. The circles and letters have the same signification as those in the brain of the monkey, fig. 64.

pital and temporo-sphenoidal lobes with each other, and are, therefore, termed by Ecker the occipito-temporal convolutions

Figure 10. In this two-page spread from *Functions of the Brain*, 1876, David Ferrier claims that the sensory motor systems of macaque monkeys and human beings are functionally identical. The drawing on the right shows the convolutions of the monkey brain, with its functional localizations experimentally derived. On the left is a schematic diagram of the human cortex onto which Ferrier has superimposed the simian neural map. The caption reads, "The circles and letters have the same signification as those in the brain of the monkey"—one of the most radical statements of animal-human continuity in the history of science.

shaped by the experiences of pain and pleasure. Pleasure encouraged behavior; pain inhibited it. In this respect, all sentient beings were alike.

Having outlined the stimulus-response psychology of dogs, Ferrier then devoted the final chapter to monkeys and humans. The chapter opened with a brief précis of the experiments on Mary Rafferty, admitting that they were "not to be commended" but concluding that they nonetheless revealed a "great fact"—the brains of monkeys and humans responded in the same way to stimulation.[66] The next four pages displayed Ferrier's maps of the sensory motor cortex of the macaque monkey next to their human counterparts. The human neural map was constructed by transposing every experimentally derived point on the monkey cortex onto a diagram of the convolutions of a human brain. Ferrier did have the grace to acknowledge that "there is nothing in man to correspond with the . . . tail in the monkey," but in every other respect monkey and man were assumed to be identical.[67] This statement of animal-human continuity, far more radical than anything to be found in Darwin, crossed an important threshold in the evolving ethics of animal and human experimentation. In Ferrier's book, charity patient Mary Rafferty and the macaque monkeys in his laboratory completely overlapped—in both their neuroanatomy and their ethical status as experimental primates.

With the publication of *The Functions of the Brain*, Ferrier's career was made. He was elected to the Royal Society in June 1876 and to the Royal College of Physicians the following year. In 1878 he cofounded the journal *Brain* as a continuation of the work begun in the *West Riding Lunatic Asylum Medical Reports*. In 1881, at an international medical congress in London, at the peak of his fame, he scored a very public victory in a theatrical showdown with Friedrich Goltz, a German colleague who opposed the idea of localization of brain function. It was the crowning moment in his career.

THE SEVENTH INTERNATIONAL MEDICAL CONGRESS OF 1881

Previous international medical congresses had been held in Paris, Florence, Vienna, Brussels, Geneva, and Amsterdam, but the seventh, held in London in 1881, was an unprecedentedly elaborate affair, probably the largest scientific conference that had ever taken place at that date. Participants included such scientific celebrities as Louis Pasteur, Jean-Martin Charcot, Rudolf Virchow, Joseph Lister, and Thomas Henry Huxley. English, French, and German were the official languages, and the proceedings were published in all three tongues as the Congress went along. It was a weeklong festival of science, "marked by an unusually abundant series of banquets, receptions and excursions," combining dazzling Victorian ingenuity, glittering aristocratic hospitality,[68] and a dash of lowbrow entertainment.[69] One Gradgrindian day saw a visit to the Croydon Sewage Farm, where the visitors lunched off its richly fertilized produce, followed by tours of the Beddington Female Orphan Asylum and the Hanwell Lunatic Asylum.[70]

On the penultimate day of the gathering, the chemist Louis Pasteur gave an account of his work with microbes.[71] He had just completed a demonstration of the power of his anthrax "vaccine" (named in honor of Edward Jenner's cowpox inoculation against smallpox) at a farm in

France. The participants in the Congress lost no time in exploring the implications of Pasteur's breakthrough, and paper after paper extolled the relation between science and medicine. Alongside the glorious therapeutic horizons opened up by germ theory, however, came a sharper spur to the development of a philosophy of scientific medicine. The opponents of animal experimentation were on the march.

The antivivisection movement in Britain had been gaining momentum since the mid-1870s. By the end of the decade there were a nearly a dozen organizations dedicated to the cause, many with their own newsletters or journals. In 1875 the movement had succeeded in getting a hearing in the House of Lords, and between July and December of that year a Royal Commission investigated the rights and wrongs of the issue, recommending limitations on animal experimentation. After three months of parliamentary debate and intensive lobbying by the scientific community, a watered-down version was passed into law, known as the 1876 Cruelty to Animals Act.

Even though the legislation was only a partial victory, it proved inspiring to antivivisectionists elsewhere in Europe, and by 1881 just about everyone at the Congress had a reason to feel embattled. Delegate after delegate made speeches mixing the triumphant history of medical progress with a rousing defense of vivisection. It was this issue that sealed David Ferrier's reputation as one of the stars of physiological science. On the second day of the Congress, the showdown between him and Goltz turned into a carefully staged demonstration of the value of animal research. The stakes were high. Localizationists like Ferrier took a brutally materialistic approach to psychology; holists such as Goltz advocated for a vision of a unified mind and for free will. Whoever won this debate would help set the moral and spiritual agenda for neurology going forward.

For all their far-reaching differences, Goltz and Ferrier were equally loathed by animal welfare campaigners.[72] To the antivivisectionists, lesion experiments on the brains of animals were particularly grotesque examples of scientific hubris. The fact that neurologists disagreed about the significance of such experiments seemed to render the whole enterprise pointless as well as cruel. The staging of a showdown between the two vivisectionists therefore held out a sweet promise: with one crucial experiment the scientists might counter the claim that research on the brains of animals represented a heinous combination of sadism and futility.

The demonstration was held in three parts. First Goltz and Ferrier outlined their arguments. Goltz described the surgeries that he had done on a dog and claimed that it had remained unimpaired in any specific function but had suffered a uniform blunting of all its abilities, alongside a marked reduction of its general intelligence.[73] From this he concluded that the brain operated as one unitary organ of thought, the material equivalent of the seat of the soul. In his response, Ferrier adroitly side-stepped his opponent. Instead of taking issue with any of Goltz's facts and conclusions, he conceded them with regard to the brain of a dog. Cortical localization, he argued, was a peculiarity of the higher animals, especially primates. He concluded with a summary of the surgeries that had been performed on two macaque monkeys, which he promised to exhibit alongside Goltz's dog that afternoon.[74]

After lunch, the audience watched a performance by the three vivisected animals.[75] First Goltz introduced his dog. It was normal in all respects, he declared, except that it had become diffusely less intelligent. He blew cigar smoke in its face to demonstrate that it did not turn away. He confided that it had happily eaten dog meat earlier in the trip. He shook a fist at it to show that it did not flinch or cringe. The operation on its brain had made the creature stupid, Goltz argued, turning it into the canine equivalent of a cannibal, madman, or drug addict, but had left it basically functionally intact.

Ferrier then brought his monkeys onto the stage. He explained again that he had removed the motor area controlling the left arm and leg in one of the creatures, and had cut away both auditory centers in the other. As the first monkey limped about, awkwardly accepting food from Ferrier with only his right hand, the great French psychiatrist Jean-Martin Charcot was so struck by its resemblance to hemiplegics at his asylum in Paris that he was heard to exclaim "It is a patient!"[76] Ferrier then shot off a series of percussion caps behind the two animals, showing that the other monkey was profoundly deaf.

All three animals were then euthanized, and a committee of delegates was deputed to examine their brains. The result was a victory for Ferrier. The lesions in the monkeys were exactly where he had suggested, whereas Goltz's dog's cortex was found to be much more intact than he had promised. On a conceptual level, this was a victory for British neurology and utilitarian psychology, but on a practical level, it was also a triumph for animal experimentation across the whole medical world. An experiment had been performed that had succeeded in resolving a profound question about the structure of the brain. Scientific cruelty was justified by the scope and novelty of the payoff.

Abridged and garbled reports of the performance found their way into the *British Medical Journal* and the *Lancet*. Both short reports asserted that the monkeys had been operated on by Ferrier, whereas the surgeries had actually been performed by Ferrier's colleague Gerald Yeo. The articles were read by an antivivisectionist, who went to look up the exact provisions of Ferrier's animal experimentation license and found that he lacked one altogether. Her animal rights organization, the Victoria Street Society, brought a prosecution. On November 18, 1881, only three months after his triumph at the Congress, Ferrier found himself in Bow Street Criminal Court, accused, under the terms of the Animal Cruelty Act, of practicing vivisection without a license.

The courtroom was packed with leading men of science, anxious to see the outcome. The charges were easy to answer. Yeo had done the surgery, and he was in possession of the requisite license. The prosecution counsel regrouped, and tried to argue that the cruelty had extended to the whole performance at the Congress, up to and including the autopsy of the animal subjects. The judge waved this away as absurd and dismissed the case.[77]

The failure of this high-profile prosecution proved to be a turning point for the antivivisection movement. If Ferrier could not be stopped with the Cruelty to Animals Act, then working through legal channels was clearly useless, the activists reasoned. On the other side, Ferrier's work on cerebral localization became the centerpiece of the medical establishment's campaign to persuade the public of the utility of animal experimentation. A scant few weeks after his appearance in court, the *British Medical Journal* published a long article detailing all the cases in which his work had already paid therapeutic dividends. His human neural maps were, according to the anonymous author, "as invaluable as a chart of an unknown region would be to an explorer."[78]

The list of cases presented in the *BMJ* article seems an impressive roster of achievements, but upon closer inspection it becomes clear that none of them are truly attributable to Ferrier's work. Most of the case histories are ones in which the patient suffered a fall or other blow to the head, and so Ferrier's maps were not required to identify the location of the injury. One of the examples concerned a man who was struck on the left side of the head with a stone and lost the power of speech. This was diagnosed as an injury to his cerebral speech center, which had been localized to the third left frontal convolution of the brain by the French anthropologist and physician Paul Broca in the 1860s. Broca had identified the speech center on the basis of autopsy material from his aphasic patients. No amount of animal experimentation, after all, could have revealed the location of a uniquely human attribute.

The most problematic case concerned one of Ferrier's own patients, a seven-year-old who suffered a blow to the head with a poker. A year later she began to suffer from seizures:

> Dr. Ferrier was asked to see the child in consultation; tenderness was found over the right parietal region, with loss of power in the left hand, and indistinct utterance from loss of muscular power in the lips. Trephining was decided upon, and Dr. Ferrier pointed out that the seat for trephining should be rather low down, to correspond to the centres in the brain for the arms and lips, which seemed affected.[79]

In this case, the animal research might have been positively misleading. The macaque's sensory motor cortex is more widely spread over the brain than that of the human. Ferrier's suggestion that the "trephining" of the cerebral center for the arms and lips start "rather low down" was based on his unwarranted assumption that his map of the monkey's brain could be extrapolated to humans. The claims made on behalf of his neural maps, in other words, represent a perilously premature attempt to justify the animal research on the grounds of immediate utility to humans.

In the mid-1880s Ferrier's colleagues embarked on a series of experimental brain surgeries based loosely on his map of the cortex. In 1884 a surgeon opened up the skull of a patient at the Queen Square Hospital and removed a tumor. Before the case was even reported in the medical press, it was the subject of a letter to the *Times* arguing that the patient "owes his life to Ferrier's experiments, without which it would not have been possible to localize his malady or attempt its removal." The man died a month later of meningitis, but his case served the medical profession as a "living monument of the value of vivisection."[80] Other cases followed, each of which was lauded as a vindication of Ferrier's research. In this way, the medical profession was able to hit back against the antivivisectionists with a cost-benefit analysis weighing animal suffering against human disease.

The medical establishment could have easily selected animal experimentation that was more directly linked to therapeutic results. Yeo's perfecting of antiseptic technique on monkey's brain tissue, for example, was more important in preparing the way for safe neurosurgery than Ferrier's localization of function. But it was precisely its apparent distance from immediate therapeutic outcomes that made Ferrier's work useful as a propaganda tool. If the researchers could defend the utility of *these* experiments—speculative, controversial, philosophically significant, and undergirded by soul-denying materialism—then surely they could make the argument that *all* vivisectionists should be left alone to pursue their questions in their own way without interference.

Far from damaging his reputation, Ferrier's criminal trial secured his scientific immortality. It also sealed the terms for Anglo-American medical research for most of the twentieth century. With touching fidelity, this version of the significance of the Goltz-Ferrier debate has survived to the present day almost unaltered. To take just one fairly recent example among many, a 2000 issue of *Neurology* claims that "Ferrier's dramatic demonstration of the effects produced by localized lesions in macaques triumphed over Goltz's unitary view of brain function, providing a major impetus for the subsequent successful development of neurologic surgery."[81] Primate research was therapeutically useful because the brains of man and monkey were fundamentally the same, but when it came to the moral justification of the research, the relief of human suffering weighed almost infinitely against the claims of any other creature, monkeys included.

These legal and ethical terms were ones that Bentham would have found most troubling. Bentham extended moral consideration to all sentient beings. In his immortal formulation, still beloved of animal rights activists, "the question is not 'can they *talk*?' or 'can they *reason*?' but 'can they *suffer*?'"[82] For him, the continuity between humans and animals ruled out cruelty to animals. Ferrier took one tentative but significant step in the other direction. In his moral universe, the fact that humans were animals opened up a space for "human vivisection," and he cautiously endorsed the experiment that killed an Irish housemaid. Ultimately, Mary Rafferty's death was justified on the grounds of its value to future neurosurgical patients, but her tragic predicament—as a feeble-minded charity case—meant that she was judged to have little to lose even under the terms of a utilitarian calculus that weighed human suffering so heavily. Animal welfare activists were outraged, but they lost the battle, and the paradoxes would not become visible again until the turmoil of the 1960s.

UTILITY AND SYMPATHY

Herbert Spencer may have been Ferrier's guru, but he was not the only prophet of evolution to wrestle with utilitarianism and pleasure-pain psychology. Charles Darwin's engagement with utilitarianism was just as long and deep as Spencer's—and just as critical. In 1838 Darwin had his famous eureka moment when he picked up Malthus's *Essay on the Principle of Population*, "for amusement," and lighted on natural selection as the mechanism by which organisms adapted to new environments and through which species changed.[83] A few months later, in September 1838, he jotted down some thoughts on William Paley's *Principles of Moral and Political Philosophy*, which he would have studied for his BA exam at Cambridge. This prompted him to ponder the vexed question of how to account for altruism in biological terms. What was the relation between evolutionary utility—to the individual, to the tribe, to the species—and utilitarianism? In his notebooks, under the heading "On Law of Utility," he proposed that the social instincts of the individual coincided with the greatest happiness principle because "only that which is beneficial to [the] race, will have reoccurred," a restatement, in evolutionary terms, of David Hume's musings on the utility of virtue of a century earlier.[84]

In the late 1860s, after the success of *The Origin of Species* had emboldened him to tackle the thorny question of his own species, Darwin started wrestling again with evolutionary ethics. By this time he had joined the tradition of British happiness theorists from Adam Smith to Herbert Spencer who rejected utilitarianism as the ultimate moral standard, on the grounds that calculating the consequences of every action

Figure 11. "Mania" and "Melancholia," from Charles Darwin's *Expression of the Emotions in Man and Animals*, 1872. These photographs of patients of the panoptic West Riding Pauper Lunatic Asylum represent the full emergence of the utilitarian self in the image of the workhouse inmate. The Hobbesian twins Felicity and Misery have now matured into medical diagnoses. Darwin's book also features many photographs of people having their facial muscles electrically stimulated to produce the supposedly universal expressions of emotional states. In a similar vein to Ferrier's *Functions of the Brain*, Darwin's *Expression of the Emotions* reveals man to be a primate with an electrical brain.

did not accord well with our social instincts. In *The Descent of Man*, published in 1871, he suggested that the "greatest happiness principle" was just a "secondary guide" to human moral action, while the "social instincts, including sympathy" served as the "primary impulse."[85]

Hutcheson, Hume, Priestley, and Paley had all assumed that the moral sense was installed in man by a benevolent deity. Adam Smith thought that the instinct of sympathy arose as a matter of necessity out of our condition as social and rational animals. Bentham, Bain, and James Mill proposed that we learn our utilitarian ethics through the experience of pain and pleasure. Darwin and Spencer reverted to the Enlightenment idea of an innate moral sense, but attributed its presence in the human

breast to the action of evolution, which selected and preserved social behaviors and sympathetic emotions that were of benefit to the tribe.

In 1872, only a year after the publication of *The Descent of Man*,
Darwin went to press with *The Expression of the Emotions in Man and
Animals*, featuring pictures of the West Riding patients and discussing
Bain and Spencer's psychological theories.[86] Darwin's musings on happiness were a delicate balancing act, somehow managing to combine the
elevated spiritual tone of his eighteenth-century predecessors with the
nihilistic explanations of his twentieth-century disciples. For Enlightenment psychologists, human happiness was a manifestation of divine
benevolence. For twentieth-century evolutionists, pleasure was just one
among many tools that had been selected because it helped us survive
and reproduce, an explanation that accounted just as well for the keen
enjoyment of rape as for the tender delights of suckling a baby. Darwin
wobbled along a narrow but elevated middle path between these two
extremes, giving happiness and pleasure rational, mechanistic explanations, but not to the extent of stripping them of their meaning and their
nobility: "We long to clasp in our arms those whom we tenderly love.
We probably owe this desire to inherited habit, in association with the
nursing and tending of our children, and with the mutual caresses of
lovers."[87] In his later autobiographical writings, Darwin appealed to the
principle of evolutionary utility to arrive at the cheerful conclusion that
"all sentient beings have been formed so as to enjoy, as a general rule,
happiness."[88] His theory of natural selection may have been pessimistic
and directionless, but he never abandoned his forebears' cheerful faith in
the ultimate victory of joy, sympathy, and goodness.

Others eventually stepped in where Darwin feared to tread, however,
speculating as to the exact workings of a utilitarian pleasure-pain system that had evolved for the purposes of survival and reproduction. The
most important evolutionary theorist of pain and pleasure was Conwy
Lloyd Morgan, professor of psychology at University College Bristol. In
an early book, *Animal Life and Intelligence*, he went so far as to declare
that without life-conserving pleasures and life-diminishing pains "it is
difficult to conceive how the evolution of conscious creatures would be
possible."[89] For Morgan, pain and pleasure constituted both the origin
and the purpose of sentience itself.

Attempting to distinguish what was learned from what was instinctive in different bird species, Morgan hatched chicks in incubators to insulate them from parental influence, and then compared their behavior with their wild-born siblings. In a little enclosure in his garden, he tried to make visible the process of learning by association with delight and misery. He threw caterpillars at chicks and observed how they learned to discriminate between the nice and the nasty ones. Nutritious caterpillars were gobbled down. The first poisonous caterpillar elicited a peck, followed by a shrinking away. The second one was rejected on sight: "Though he sees he does not seize, but shrinks without seizing," Morgan punned. In explanation, he gave a neurological account of pain and pleasure and their effect on behavior:

> Certain stimuli call forth disturbances, probably in the cortex of the brain, the result of which is the inhibition of activities leading to repetition of these stimuli; certain others call forth cortical disturbances, the result of which is the augmentation of the activities which lead to their repetition. The accompaniment in consciousness of the former we call unpleasant or painful; the accompaniment in consciousness of the latter we call pleasurable.[90]

In his passage about whipping a dog to curb its appetite, Ferrier had dealt only with the pain side of the equation. Conwy Lloyd Morgan added in pleasure. Attraction and repulsion, desire and avoidance, appetite and aversion: the fundamental Hobbesian dichotomies were beginning to be systematized into a science of animal behavior. The next chapter will explore the transatlantic passage of this utilitarian psychology to Harvard University, where it was appropriated as an American invention, stripped of its remaining humanist aspects, and turned into a highly successful scientific movement.

5

In Which We Wonder Who Is Crazy

ANIMALS STUDIED BY AMERICANS

Conwy Lloyd Morgan's 1896 book *Habit and Instinct* described how his new-hatched plover chicks would lie bedraggled in the incubator drawer, tiny necks stretched out, until they started to stir. He recounted how his "little friends" shrank from his touch at first, but once they got over their fear would come running into his hands and "poke out their little heads confidingly between my fingers." He described their "consummate impertinence" in pecking the toes of his endlessly patient terrier, Tony. He talked of their scrambling to get onto their "favourite raft, made out of an old cigar box." He quoted at length from a travel writer, who portrayed a line of baby pheasants—"tiny, fluffy balls of down, not many days out of the shell"—"struggling manfully" behind "an old bird . . . moving her head jauntily as if with the conscious pride of motherhood."[1]

It is therefore somewhat ironic that Morgan is best remembered for "Morgan's Canon," his exhortation against anthropomorphism. "In no case may we interpret an action as the outcome of a higher psychical faculty, if it can be interpreted as the outcome of the exercise of one that stands lower in the psychological scale," he insisted in 1894.[2] Scientists of animal behavior, in other words, must interpret every action according to the lowest, most basic psychological faculty that could possibly account for it. But he described life among his baby birds with such detailed tenderness that he seemed incapable of obeying his own commandment. For all his brave words about scientific objectivity, Morgan's stories of his fluffy infants learning to love their human parent were simply too fond, and his chicks ended up just as charming and anthropomorphic as any loyal dog, questing pea tendril, or affronted primate in Darwin.

Habit and Instinct was based on a series of lectures that Morgan delivered on a tour of American universities. His Boston lecture was attended by a Harvard graduate student named Edward Thorndike, who seized on Morgan's basic framework, boiled it down to a few central precepts, and set about formalizing it into the experimental protocols of American behaviorism.

Thorndike was well primed to receive the baton of utilitarian psychology. His university education had been completely steeped in the British literature on pain and pleasure. In 1895 he wrote an undergraduate thesis at Wesleyan University entitled "A Review and Criticism of Spencer's *Data of Ethics*."[3] The following year, he went on to pursue graduate study in literature at Harvard, fell under the spell of the professor of psychology William James, and switched over to his new idol's field. James's curriculum included Alexander Bain, John Stuart Mill, Herbert Spencer, and Conwy Lloyd Morgan.

Shortly after attending Morgan's Harvard lecture, Thorndike resolved to build an animal behavior laboratory of his own, no doubt convinced he could make a better fist of it than the sentimental Englishman. Because animals were not allowed in the psychology department, William James allowed him to rig up some equipment in the cellar of his grand wood-frame house. Down in this dusty cave, Thorndike embarked on his first series of experiments on cats, dogs, and chicks.[4] (One wonders if the hammering and yowling ever came to the ears of the distinguished family above.) In 1897 he transferred to Columbia University, where he continued the research for one more year. After writing his experiments up in a boldly polemical dissertation, he was awarded his doctorate at the age of twenty-four.

It turned out that Thorndike had a peculiar talent for ensuring that animals displayed no intellectual capacities whatsoever. "Most of the books," he snorted, "do not give us a psychology but rather a *eulogy* of animals. They have all been about animal *intelligence*, never about animal *stupidity*."[5] Particularly irritated by descriptions of cherished cats ingeniously turning doorknobs and springing latches, he announced that he had designed his experiments specifically to refute such accounts of feline brilliance.

Thorndike explained that he had constructed "puzzle boxes"—little wooden cages with doors that opened when pressure was put on a simple pulley-and-latch mechanism. Outside the door he dangled a piece of fish; inside the box he placed a starving cat. The cat would start out clawing and scratching at the walls of the box at random. If it opened the door of the puzzle box by accident, the pleasurable result—gobbling down the fish—conditioned it to repeat the random action. Gradually, through repetition of the tasty outcome, the cat "learned" to open the box. From this Thorndike concluded that that *all* animals surmounted *all* challenges in this manner—not by any intelligent process, but just through the training of their reflexes through satisfaction and frustration. With no powers of reason, memory, or imitation, they mastered the world solely through the dull repetition of pains and pleasures.

The essence of Thorndike's argument was contained in a graph. He plotted how long it took the cat to open the latch in a series of repeated trials, and then claimed that resulting line could be resolved into a smooth curve. There was no one moment when the creature learned the trick. "The gradual slope of the time curve," he declared, "shows the absence of reasoning."[6] With this evidence, Thorndike demonstrated to his absolute satisfaction that the animals mastered the task, not by anything resembling intelligence, but by a process he called "stamping in" and "stamping out." There was no point at which the cats "solved" the puzzle; they just absorbed the lessons of life through ceaseless repetition of pleasure and pain. "Gradually," he explained, "all the other non-successful impulses will be stamped out and the particular impulse leading to the successful act will be stamped in by the resulting pleasure, until, after many trials, the cat will, when put in the box, immediately claw the button or loop in a definite way."[7]

In 1911 Thorndike reprinted his dissertation with a collection of other papers. The penultimate chapter gave a name to his theory, the "Law of Effect": "Of several responses made to the same situation, those which are accompanied or closely followed by satisfaction to the animal will,

other things being equal . . . be more likely to recur. . . . Those which
are accompanied . . . by discomfort to the animal . . . will be less likely
to recur."[8] Up until the publication of this book, the principle connecting pain, pleasure, and learning was known in the United States as the
"Spencer-Bain theory."[9] Thereafter it was known as "Thorndike's Law of
Effect." Many years later, as Thorndike's fame grew, one of his disgrun-
tled colleagues was moved to publish an article taking him to task for his
failure to acknowledge his indebtedness to his British predecessors.[10] He
never yielded his claim of originality, however. With Thorndike's cheerful
appropriation of four decades of British research, pleasure-pain psychology became, in the eyes of the world, an American invention.

Eventually tiring of the work with chicks, cats, and dogs, Thorndike
turned his attention to the human animal. He got a professorship at Columbia Teachers College and devoted the rest of his life to the development of a scientific educational psychology. Applying his Law of Effect
to the classroom, he found that reward, stimulation, and interest were
better guarantees of learning than drills and punishments. A series of
studies undertaken in the 1920s and 1930s with adult learners showed
that the positive pole of the Law of Effect was far more potent than the
negative one.[11] Reward, in other words, was the main motivator, with
punishment coming in a very long second. Energetic, collegial, and pro-
lific, Thorndike became one of America's most versatile policy advisors,
available for committee work on issues ranging from ventilation to movie
censorship to intelligence testing. A glorious career was crowned with the
presidency of the American Association for the Advancement of Science.

In 1935 a party was thrown at the AAAS on the occasion of Thorndike's retirement as president. His speech was entitled "Science and
Values." Arguing that "the topic is important for workers in all sciences,"
he claimed that he had discovered the true foundation for ethics: "Judgments of value are simply one sort of judgments of fact, distinguished
from the rest by two characteristics: they concern consequences. These
are consequences to the wants of sentient beings." In one pithy phrase—
"consequences to the wants of sentient beings"—Thorndike summed
up the essence of Bentham's moral philosophy. For Thorndike, as for
Bentham, all that mattered in ethics was the effect of our actions on the
pains and pleasures of feeling creatures: "Values appear in the world
when certain forms of preference appear, when certain animals like or

dislike, enjoy or suffer, are contented or unhappy, or feel pleasures or pains." As usual, Thorndike did not acknowledge that anyone had arrived at this conclusion before him, observing instead that moral philosophy had made no discernible progress in two and a half millennia: "Aristotle's solutions seem as good as Hegel's."[12]

Thorndike had succeeded in founding utilitarian psychology anew, relieved of all the humanistic baggage that had weighed it down in nineteenth-century Britain. Back in the mother country, the threads that tied Bain and Spencer back to Bentham and all the pressing questions about political reform were still intact. In the United States, the Spencer-Bain theory was renamed the Law of Effect and reinvented as pure, value-free, laboratory science, with no connection to politics and history. In 1935 this process came full circle, when Thorndike gave his speech presenting utilitarianism as merely the logical derivation of his experiments in pleasure-pain psychology. From that time forward, the utilitarian values embedded in behaviorism would enjoy the imprimatur of value-free science.

For all his assiduous self-promotion, Thorndike did not, in the final analysis, enter the halls of scientific immortality. His claim to be the father of behaviorism was thoroughly eclipsed by that of John B. Watson, surely one of the most unlikable characters ever to make his mark on science. Born into a hardscrabble life in South Carolina, where "Nigger fighting" became one of his favorite hobbies, Watson was fiercely ambitious for social status and professional advancement.[13] After obtaining an undergraduate degree at a local college, he wrote to the president of the University of Chicago declaring himself "poor" but "earnest" and "very anxious . . . to do advanced work at real university."[14] His earnest brashness paid off; he was given a place at Chicago, and in the fall of 1900 he arrived at the Rockefeller Gothic campus with fifty dollars in his pocket.

At Chicago, Watson introduced the white rat into the psychology of learning, systematically removing its different sense organs in order to see what effect each deficit might have on its ability to solve a maze. He got into trouble with animal welfare advocates, but his ruthless pursuit of reductive explanations for animal behavior and his flamboyant debating style earned him a fierce reputation, and he was widely courted for jobs (including by Thorndike at Columbia). He chose Johns Hopkins and went there as a professor of psychology in 1908.

In 1913 Watson gave a series of unexpectedly popular lectures in New York City, beginning with a manifesto entitled "Psychology as the Behaviorist Views It." Declaring that the goal of psychology was the "prediction and control of behavior," he insisted that the discipline treat humans and animals alike as stimulus-response mechanisms.[15] The only way to place psychology on a truly scientific footing would be to recognize "no dividing line between man and brute."[16] In effect, Watson exhorted scientists to stop being anthropomorphic about *humans*.

Watson's promise of rational mastery of human action struck a chord. This was the time when scientific management was being touted as a solution to the dislocations of industrialization, and the avatars of efficiency embraced the behaviorist creed with enthusiasm. After the Great War, Watson perfectly captured the Progressive Era zeitgeist by founding the Industrial Services Corporation to provide psychological evaluations of workers and increase their output per hour. In his large and capable hands, "behaviorist" psychology became *the* bastion of antihumanism at the center of the human sciences.

The technocrats who funded Watson were not put off his creed even by views on child-rearing that now seem extraordinarily callous. Himself the father of two boys, Watson advocated that children should not be touched by their parents beyond a firm handshake. At one point he suggested that caregivers might "arrange a table containing interesting but not to be touched objects with electrical wires so that an electrical shock is given when the table to be avoided is touched." This idea won him a contract from the Rockefeller Foundation to develop "techniques that would reduce child rearing to standardized formulae."[17] After Watson's death, one of his sons committed suicide.

A most unlikely perpetuator of the idea that Thorndike and Watson invented animal behavior science was the British philosopher Bertrand Russell. When Russell read the behaviorists, he judged their work compatible with his own ambition to rebuild all knowledge in strict accordance with logic and the facts. In a 1927 book written for an American audience, he assigned behaviorism a core place in his philosophical framework, announcing that "the study of learning in animals" may be "regarded as beginning with Thorndike's *Animal Intelligence*, which was published in 1911."[18]

Russell's obliteration of the British tradition of animal experimenta-

Figure 12. Conwy Lloyd Morgan. In 1896 Morgan brought British animal behavior science to the United States, only to suffer the eclipse of his reputation by his transatlantic imitators Edward Thorndike, John Watson, and B. F. Skinner. Morgan represents the apogee of a long tradition of British research into the motivating force of pain and pleasure. His work is distinguished by its warmth, humanism, and subtlety, characteristics notably absent from the experiments of his American disciples.

tion cannot be attributed to ignorance. When he was young, his father, a disciple and friend of John Stuart Mill, hired as a tutor one D. A. Spalding, a protégé of Alexander Bain who raised chickens all over the house in order to study their habits and instincts, in the same vein as Conwy Lloyd Morgan. Here is Russell's description of Spalding in his autobiography:

> He was a Darwinian, and was engaged in studying the instincts of chickens, which, to facilitate his studies, were allowed to wreak havoc in every room in the house. He himself was in an advanced stage of consumption

and died not long after my father. Apparently on grounds of pure theory, my father and mother decided that although he ought to remain childless on account of his tuberculosis, it was unfair to expect him to be celibate. My mother therefore allowed him to live with her, though I know of no evidence that she derived any pleasure from doing so.[19]

Given the fact that an English animal experimentalist was sleeping with his mother, Russell's assertion of the priority of American scientists over British ones is a bit baffling.

Russell did, at least, have some fun at the foreigners' expense. Comparing Thorndike's work with a German scientist's research on monkeys, he drily remarked that the experiments "displayed the national characteristics of the observer. Animals studied by Americans rush about frantically, with an incredible display of hustle and pep, and at last achieve the desired result by chance. Animals observed by Germans sit still and think, and at last evolve the solution out of their inner consciousness."[20] About the animals observed by his own compatriots—all those thousands of experimental creatures raised tenderly in drawing rooms, gardens, and greenhouses in the decades since Darwin's *Origin of Species*—Russell's legendary wit was silent.

OPERANT CONDITIONING

Just as Morgan was obscured by Thorndike, and Thorndike was outshone by Watson, so Watson's handsome face eventually bore the bootprint of his successor, Burrhus Frederic Skinner. B. F. Skinner combined Thorndike's absolute trust in the power of reward with Watson's resolute refusal to be anthropomorphic about *Homo sapiens*. With this, the utilitarian aspects of behaviorism really came to the fore. Skinner's mechanistic psychology more or less entailed extreme utilitarian ethics. If humans are nothing more than automata, pushed and pulled by pain and pleasure, it follows that worrying about freedom, dignity, or autonomy is simply a waste of time. Rights are nothing more than empty words. Pain and pleasure, by contrast, are inarguable physiological truths. Totting up the pleasure side of the ledger and subtracting the pain side is self-evidently the only ethic with any pretensions to rationality.

Every bit as bumptious about the world-changing potential of his psychological insight as Bentham, Skinner was born in 1904 and grew up in Susquehanna, Pennsylvania, which just happens to be the sleepy farming community in which Joseph Priestley had lived out the last years of the eighteenth century in political exile. After completing an undergraduate degree in English literature, he moved back in with his parents, set up a desk in the attic, and tried his hand at writing literary fiction, a disillusioning experience that he later dubbed his "Dark Year." It was in his capacity as a failed artist that Skinner turned to science. Consoled by an article that judged Ivan Pavlov a more important figure than Leo Tolstoy, he enrolled in as a graduate student in psychology at Harvard.

Of those first years at Harvard, Skinner recalled that he was confirmed in his choice of discipline not so much by what he was learning in his classes "as by the machine shop in Emerson Hall." An adept tinkerer and inventor since childhood, he developed a battery of technologies for measuring reward-seeking behavior, ending up with an "experimental arrangement not unlike the famous problem boxes of Edward L. Thorndike."[21] These "Skinner boxes" were rigged up with a mechanism that enabled the animal to get a food reward—a lever for a rat to press or a disk for a pigeon to peck. Skinner figured out how to make the rate at which rats helped themselves to food pellets self-recording, and was able to show the mathematical regularities in the resulting graphs. As with Thorndike, the argument rested on the smoothness of the curve. Because the rate at which the rat helped itself to food was regular, the creature could not be said to be making a choice. "In my experiment the only thing the rat had to do was to decide: to eat or not to eat. But the decision was made *for* the rat . . . by some orderly physiological change."[22] His graphs sufficed to satisfy Skinner that free will was an illusion, in rats and humans alike.

After obtaining his PhD, Skinner carried on at Harvard as a research fellow, where he began to develop his own vocabulary for pleasure-pain psychology. First he introduced the word "reinforcement" for a pleasurable outcome of an action that increased the likelihood of the action being repeated—for example, the tasty food pellet that rattled into a little trough when the rat pressed the lever.[23] Later he coined the word "operant" for behavior that "acts upon the environment in such a way that a reinforcing stimulus is produced"—such as the act of pressing the lever and getting the food reward.[24] "Operant conditioning" was the process by which behaviors—operants—were either reinforced by pleasure or extinguished by pain. His first book, *The Behavior of Organisms*, was published in 1938, by which time he was married, with a baby daughter, and had secured an assistant professorship at Minnesota. When one review accused him of downplaying his intellectual debts, Skinner wrote to Thorndike: "It has always been obvious that I was merely carrying on your puzzle-box experiments, but it never occurred to me to remind my readers of the fact."[25]

Skinner did, however, refine and elaborate Thorndike's Law of Effect. One of his most important discoveries was made by accident, when his food dispenser jammed and he found that the rat pressed the lever even more compulsively when it suddenly stopped producing rewards. It was a Friday afternoon, and there was no one to tell. "All that weekend I crossed streets with particular care and avoided all unnecessary risks to protect my discovery from loss through my death."[26] Skinner was able to systematize this into a "schedule of reinforcement" in which rewards coming more rarely made the pursuit of them increasingly compulsive. Eventually he worked out how to stretch the reward ratio a little at a time, inducing pigeons to keep pecking thousands of times for a single grain of corn, like a gambler at a cannily programmed slot machine.[27] With this technique, Skinner could get a rat or a pigeon to do just about anything, and his displays of animal behavior began to attract a cadre of followers for whom devising reductive explanations for human action became a badge of scientific toughness.

Ambitious and intellectually single-minded, Skinner had never taken much interest in politics. "Roosevelt," he remarked, "slipped into office almost unremarked by me." He knew enough to be "against Hitler," however, and the German invasion of Norway and Denmark did get his attention.[28] In April 1940, traveling by train through the Midwest, thinking about how the Nazis had turned airplanes into bombers, he noticed a flock of birds lifting and wheeling in formation and had a thought: "Suddenly I saw them as 'devices' with excellent vision and extraordinary maneuverability. Could they not guide a missile?"[29] "Project Pigeon" received funding from the National Defense Research Committee but never went beyond prototype stage.[30]

The "Bird's Eye Bomb" was Skinner's Panopticon—a cherished design that came to naught but that marked a sharp escalation in his worldly ambitions. He began to conceive of his behavior control techniques as applicable to human life outside the laboratory, and after the war he issued a series of manifestos urging social reform by means of positive reinforcement. The first of these was a utopian novel, which he completed in seven weeks in 1945, while on a Guggenheim Fellowship. After the Guggenheim year was over, he became the new chair of the psychology department at Indiana University, but he stayed for only three years, moving in 1948 to Harvard.

At first, no publishers were interested in Skinner's novel, until eventually Macmillan rather grudgingly brought it out, in 1948, under the title *Walden Two*. The story is told from the point of view of a burnt-out professor of sociology who discovers that one of his former graduate students, a "queer duck" named Frazier, has set up an intentional community. Frazier's "Walden Two" turns out to be a couple of short bus rides away, and so the professor sets off with a band of explorers, including a cranky professor of philosophy, to see if it is, in fact, possible to craft a viable alternative to capitalism.

Frazier meets the group at the bus stop and drives them into a beautiful valley, dotted with unassuming earth-colored buildings. Over the next few days, he leads them around the dining rooms, kitchens, manufacturing sheds, schools, farms, gardens, libraries, and theater, explaining how all the labor-saving devices and systems allow the members to focus on "art, science, play, the exercise of skills, the satisfaction of curiosities, the conquest of nature—the conquest of man himself, but never of other men . . . leisure without slavery."[31] Naturally, everyone there is happy, healthy, creative, cooperative, and kind. Most of them are also educated, intelligent, and high-minded, with tastes running to German classical music and good poetry. The women are beautiful in an understated way. This bucolic vision has been likened to Skinner's own "small town, middle-class Protestant background."[32]

One notable exception to the serenity of the communards is Frazier himself, Skinner's alter ego and the book's most (some say only) compelling character, who turns out to be a socially awkward, volatile megalomaniac. In one scene, he says to the professor, "You think I'm conceited, aggressive, tactless, selfish . . . completely insensitive to my effect on others, except when the effect is calculated . . . my motives are ulterior and devious, my emotions warped." He then smashes a clay tile against

Figure 13. B. F. Skinner with pigeon. Skinner sits in a machine shop surrounded by boxes and wires, wearing an apron, as befits his status as an adept technician who designed and built some of the most canonical laboratory tools in the history of psychology. The picture was taken in 1948, at a high point in his career. He had just moved to Harvard, and *Walden Two* had recently been published. In the novel, Skinner's alter ego compares himself favorably to God, perhaps belying this image of a sober scientist concerned only with understanding pigeon and rat behavior. Courtesy of dpa picture alliance/Alamy Stock Photo.

the fireplace with the words "Can't you see? *I'm—not—a—product—of—Walden—Two!*"[33]

By staging debates between his alter ego Frazier and a series of equally argumentative professors, Skinner was free to explore behaviorist philosophy unconstrained by the conventions of scientific prose. The results are revealingly grandiose. In one episode Frazier surveys the community through binoculars, then throws himself down on the grass and admits there is a "curious similarity" between himself and God, the main dissimilarity being that Walden Two is less "disappointing" than God's creation.[34] In another he defends himself against the charge of totalitarianism: "If he takes any step which reduces the sum total of human happiness, his power is reduced by a like amount. What better check against a malevolent despotism could you ask for?"[35] With this deft move, Skinner asserted that the power of a behaviorist leader was no more or less than a measure of his rightness.

Sales of *Walden Two* were disappointing, but Skinner's visibility grew exponentially with his move to Harvard. There he taught the principles of operant conditioning using live pigeons in the classroom. To make his point against the humanistic vocabulary that he despised, Skinner trained pigeons to display what he called "synthetic social relations," getting them to play avian Ping-Pong against one another, and to peck at disks in unison in order to get food rewards. "Cooperation and competition were not traits of character," he concluded, but merely behaviors "capable of being constructed by arranging special contingencies of reinforcement."[36]

Despite the remarkable consistency with which pigeon behavior could be shaped in this way, classical behaviorists of the Watsonian school questioned the whole notion of operant conditioning, asking how a reward coming *after* an action could strengthen the likelihood of that action being repeated.[37] In 1953 the aura of mystery began to dissipate. Two postdoctoral fellows in psychology at McGill University decided to implant electrodes in the brains of rats before putting them in Skinner boxes. They were testing the aversive effects of direct electrical stimulation, but a misplaced electrode seemed to have the opposite effect. Realizing that they had stumbled upon a possible "pleasure center," the pair concocted a new device. They connected a lever in the cage to an electrode implanted in the rat's pleasure center, and measured the rate at which the animal pressed the lever.

It turned out that the poor little beast pressed the lever compulsively. The researchers concluded that "the control exercised over the animal's behavior by means of this reward is extreme, possibly exceeding that exercised by any other reward previously used in animal experimentation."[38] With their serendipitous discovery of an area of the brain that rats seemed to want—desperately—to stimulate, the two Canadians supplied operant conditioning with a plausible biological underpinning. The experiment became one of the great animal fables of psychological research, with a lever-pressing rat standing as a potent image of pleasure-seeking and addiction in humans.

THE EROS OF DEMOCRACY

Unfortunately for Skinner, World War II gave some distinctly unbehaviorist values a new lease on life.[39] The movement for informed consent in medical ethics had an exact parallel in psychiatry, fueled by the same anxieties about fascism and totalitarianism. Starting in the late 1940s, scholars began to study the psychology of democracy. At the center of these investigations was the ideal of the "democratic type"—an intelligent, open-minded, autonomous human, capable of exercising his or her psychological freedom. Behaviorism seemed to deny the very possibility of this type of psychological independence. Skinner was accordingly denounced as the prime exemplar of the antidemocratic type. With fascism looming to the right and communism to the left, inalienable rights, psychological autonomy, and personal responsibility—the very concepts that behaviorists derided as pernicious superstitions—were touted as the foundation for rebuilding a better world.[40]

In 1950 *The Authoritarian Personality* was published, a thousand-page psychoanalytic study of fascism and ethnocentrism sponsored by the American Jewish Committee. The goal was to understand the "social-psychological factors which have made it possible for the authoritarian type of man to threaten to replace the individualistic and democratic type prevalent in the past century and a half of our civilization, and of the factors by which this threat may be contained."[41] Subjects were measured on the "F scale" for how latently fascist they were, on the "E-scale" for their ethnocentrism, and on the "AS Scale" for their anti-Semitism. High scorers on any of the scales turned out to be very unpleasant people indeed, prone to criminality, superstition, delinquency, and a variety of mental illnesses. While acknowledging that "the prejudiced are the better rewarded in our society as far as external values are concerned," the book concluded that "the tolerant . . . are, basically, happier than the prejudiced. . . . If fear and destructiveness are the major emotional sources of fascism, *eros* belongs mainly to democracy."[42]

The study was criticized for its Freudianism (high scorers were seeking "sadomasochistic resolution of the Oedipus complex"[43]) and for neglecting leftist totalitarianism in favor of the far-right variety. Despite these quibbles, work continued apace on the cluster of dispositions that constituted the antidemocratic type. One researcher, for example, administered the Ethnocentrism Scale to a group of college students and then showed them a simple asymmetrical diagram. After removing the figure from view, he asked them to draw it from memory. Two weeks later he asked them to reproduce it again, and then again after four weeks. His graph of the results showed that "those high in ethnocentrism displayed an increasing tendency to reproduce the figure symmetrically." Just as the authors of *The Authoritarian Personality* had suggested, conformity not only had unpalatable political consequences, it also generated cognitive "lies and errors in vision, memory, or logic."[44]

Behaviorists soon began to come under fire for displaying just those qualities of rigidity and narrow-mindedness that characterized the authoritarian personality. A 1949 paper in *Psychological Review* accused behaviorism of having a "static, atomistic character," lacking creativity, neglecting questions of value, intensifying scientific hierarchies, and being devoid of "vitality and courage."[45] At Skinner's home institution of Harvard, the antibehaviorists founded the interdisciplinary Department of Social Relations, "identified by its interests in personality, social issues, and a commitment to the view that humans are naturally autonomous." Skinner and his mentors were left behind to rule over a psychology department defined by a rigidly deterministic perspective on human nature.[46] As the two sides entrenched, antibehaviorist critiques tended more toward the insultingly diagnostic, with an important 1959 survey of psychological science denouncing behaviorism's "attachment to a 'facile' mythology of perfection, as well as its 'autism.'"[47]

For its critics, the behaviorist universe was devoid, above all, of *meaning*. There seemed to be no place in it for concepts, judgments, values, intentions, or beliefs. The question of meaning was best exemplified, of course, by syntactical language. How could a stimulus-response analysis possibly account for the importance of word order in generating sense or nonsense? How could operant conditioning explain the proverbial newsworthiness of "man bites dog"? Taking up the gauntlet, in 1957 Skinner published an audacious treatise on *Verbal Behavior* in which he asserted that all linguistic phenomena could be analyzed in behaviorist terms: the meaning of words was a function of the reward given for a certain behavior, such as looking at a chair, uttering "chair," and being rewarded by a parent's delighted hand-clapping.

The book was received politely at first, until the linguist Noam Chomsky published a long, damning review, claiming that the theory of operant conditioning might contain valuable insights into the psychology of rats and pigeons, but extrapolating it to human symbolic language deprived it of all its precision. He advanced an alternative hypothesis, in which all languages partook of a universal grammar, produced by the unique structure of the human brain and showing marked developmental stages. The review soon began to attract more attention than the book itself. For Chomsky, Skinner's analysis could not account for meaning, could not accommodate ambiguity, and failed to explain the extraordinary ease with which human children pick up languages in the absence of anything resembling a schedule of reinforcement. In the years that followed, Chomsky helped develop a research program in psycholinguistics, devoted to probing the fundamental principles of operation of the human mind. Soon, the new field was housed in its own institution, the Harvard Center for Cognitive Studies.[48] Standing for psychological autonomy, universal human rights, and resistance to tyranny in all its forms, Chomsky felt about Skinner much as Charles Dickens had felt about Malthus a century earlier.

STUDIES IN SCHIZOPHRENIA

Meanwhile, far from the Ivy League spotlight, at Tulane University, invasive behaviorist research on human subjects was able to flourish unchallenged. Under the directorship of Robert Galbraith Heath, the Tulane Electrical Brain Stimulation Program investigated the pain and pleasure circuits of patients in the Charity Hospital in New Orleans. Implanting electrodes in the deep brain structures of schizophrenic subjects, and manipulating mood and behavior by turning the current on and off, Heath and his team enacted the most extreme performance of utilitarian psychology in the annals of neuroscience.

Heath, the son of a doctor, had obtained his medical degree from the University of Pittsburgh in 1938. Because he had specialized in neurology, he was drafted into the army during World War II, given a two-month training in psychiatry, and thrown into the deep end, treating psychic war trauma.[49] After the war he continued his psychological and neurological training at Columbia University. There he came under the influence of an émigré Hungarian psychoanalyst, Sandor Rado, who was rewriting all the themes of Freudian psychoanalysis—sex, defecation, death, dreams, anxiety, Oedipal desires, etc.—in terms of what he called "hedonic self-regulation" and evolutionary utility. Here is Rado on the joys of the table: "It is a very impressive arrangement of nature that an activity of such high utility value as ingestion should have such tremendous pleasure value for the organism."[50] Heath was spellbound, later describing Rado as "a brilliant person, extremely creative, very intuitive in his understanding," with "a more basic understanding of human behavior than any individual before or since, including his mentor Sigmund Freud."[51] It was from Rado that Heath inherited the evolutionary analysis of pain and pleasure that would define his research agenda.

When Heath arrived at Columbia, a study with patients at the Greystone Hospital in New Jersey was under way, intended to bring some scientific rigor to the assessment of neurosurgical treatments for mental illness. A group of forty-eight patients was selected and sorted into twenty-four pairs with matching psychiatric characteristics. One member of each pair was then operated on, while the other served as a control. Heath's job was the administration of the postoperative tests. The results of the study were not encouraging, but Heath was more positive than most of his colleagues, suggesting that the removal of "Brodmann areas" 9 and 10 produced lasting improvements in some schizophrenic patients. He analyzed schizophrenia as a disorder of hedonic regulation: "an inferior attempt on the part of the patient to adapt to the intense and disturbing emotions."[52] Schizophrenic delusions were just attempts to regulate aversive feelings. Feelings of failure and disappointment, for example, produced delusions of grandeur as a counterweight. If the disturbing emotions could be controlled, Heath argued, the delusions would take care of themselves.

Inspired by Rado's suggestion that the seat of the emotions was to be found in the deepest, most primitive parts of the brain, Heath began to argue that the frontal cortex was the wrong place to intervene in emotional disorders. Perhaps a better approach would be electrical stimulation of structures inaccessible to the surgeon's knife? He did not find any takers among his East Coast colleagues, but in 1949, before he had even completed his doctoral research, he was recruited by Tulane to head a new Department of Psychiatry and Neurology. He later recalled that New Orleans offered "real opportunities, because it was such a backward area . . . there was no department, but we did have this vast institution called Charity Hospital. There was a tremendous amount of clinical material."[53] The dean had high hopes for his new recruit, expressing the belief that his work "may be of the stature to be considered for the Nobel Prize," and promised him unlimited academic freedom.[54] Heath would stay at Tulane until his retirement thirty years later.

In 1950 Heath embarked on his research, aided by an interdisciplinary team of psychologists, psychiatrists, physiologists, surgeons, and clinicians. In the first round, twenty patients with "hopeless" schizophrenia were implanted with electrodes in a variety of deep brain structures.[55] Also included were three patients with intractable pain from cancer, two with psychosis associated with tuberculosis, and one with severe rheumatoid arthritis, to serve as controls. The placement of the electrodes was based on Heath's conclusion that the removal of Brodmann areas 9 and 10 produced therapeutic improvements in the Greystone patients. Since that time, he had conducted a series of animal experiments verifying that these areas were anatomically connected to the "septal region," the deep brain structure he believed was responsible for painful and pleasurable emotions. For experimental purposes, he also chose a few other sites for implantation.

In June 1952 Heath organized a three-day symposium to showcase the results. He and his team described their animal research, outlined their surgical techniques, projected films of the subjects before and after the treatment, showed graphs of their improvement, presented the case studies, and detailed the results of the postoperative tests. Heath's final assessment of the treatment was wildly affirmative. According to a "Table of Therapeutic Results," only four out of the twenty failed to show any improvement, two of whom were judged to be "technical failures" and so did not count. The rest were considered to have made progress, including five who had gone from "hopeless" before the treatment to "minimal defect" afterward:

> The outstanding changes were: ability to relate to other people, increased
> responsiveness to pleasure, gradual appearance of a sense of humor,
> and more overt expression of anxiety and ambivalence. Less negativism
> was displayed and less autistic preoccupation; everyday problems were
> approached more realistically and more interest was shown in ward
> activities. Underproductive patients became talkative. Intellectually, most
> patients seemed better motivated in performing simple test problems.[56]

Some of the invited discussants were impressed, others less so. Responses ranged from the adulatory—"The change is so dramatic and so real that one is reminded of the fairy story of 'The Sleeping Beauty'"—to the skeptical—"No sound basis has been advanced here for the assumption that schizophrenia is due to specific septal pathophysiology or that this condition is influenced by manipulations of this region."[57]

But even the most skeptical respondents congratulated Heath and his team for their courage in undertaking such a risky and speculative procedure. One attendee denounced the project's theoretical, methodological, and interpretative weaknesses before agreeing that "it is of the greatest importance that interdisciplinary work of this sort shall be carried forward not merely, though especially, in psychiatry, but in all fields of medicine."[58] Indeed, relative to the times, Heath's advocacy for the electrical stimulation program was far from extreme. In 1952 the neurologist Walter Freeman was still touring asylums all over the country, lining patients up, knocking them out with electroshock, and performing fifteen-minute icepick lobotomies, sometimes dozens at a time. The introduction of the "chemical lobotomy" in the form of antipsychotic drugs was still in the very first stages. Compared with irreversible surgery, electrical stimulation was a relatively conservative and plausible intervention.[59]

It did not remain so for long. As a result of lithium, chlorpromazine, and other drug treatments, treatment protocols changed dramatically in the decade up to 1962. Psychosurgery rates declined precipitously as a result, but Heath carried on experimenting with the electrical stimulation technique while therapeutic and ethical standards transformed around him. In some ways, his work got cruder. Inspired by the Canadian success with the pleasure centers of rats, his experiments became more straightforwardly behaviorist. Earlier, he had hoped that electrical stimulation would unscramble defective neural wiring and reverse the symptoms of schizophrenia. Later, he used the same technique in a more direct attempt to manipulate human behavior with a blast of electrical stimulation to the pleasure or pain centers of the brain.

At the beginning of the 1960s, he was once again ready to showcase his results. This time, however, he was out of step with the wider world. Now that patients could be treated for psychosis with a pill, invasive surgery of dubious therapeutic value seemed a far more reckless and irresponsible procedure. Heath also had the misfortune to preside over his second major symposium just as the electrode-wielding behavioral scientist became an emblem of Cold War malevolence, courtesy of a social psychologist at Yale by the name of Stanley Milgram.

OBEDIENCE TO AUTHORITY

The foreword to *The Authoritarian Personality*—the 1950 study of fascist psychology that brought behaviorism into disrepute—contains a startling statement: "Today the world scarcely remembers the mechanized perse-cution and extermination of millions of human beings only a short span of years away."[60] It was not until the beginning of the next decade that discussion of Nazi atrocities became widespread, spurred by the trial of Adolf Eichmann, the Nazi war criminal who was arrested in Argentina in 1960 and tried in Jerusalem. One of the people whose understand-ing of the Holocaust was crystallized by the Eichmann trial was Stanley Milgram, whose "obedience to authority" experiment fed the mounting distrust of pleasure-pain research.[61]

Milgram was the child of two Eastern European Jews who had come to the United States in the first decades of the twentieth century. In 1960 he completed a PhD dissertation probing the differences in "so-cial conformity" between Norwegians and French people. The setup was simple: it recorded how many times a subject would give an obviously false answer to a question when he or she was tricked into believing that everyone else in the study had answered that way. These experiments were cast as investigations into "national character" (Norwegians turned out to be more conformist than the French) but were also a manifestation of the wide-ranging anxieties of the 1950s about organization men, au-thoritarian personalities, and the relation of the individual to the crowd.

Sometime in the late spring of 1960, inspired by Eichmann's arrest, Milgram conceived of sinister twist on his conformity research, asking if "groups could pressure a person into . . . behaving aggressively towards another person, say by administering increasingly severe shocks to him."[62] In October and November of that year, he sent letters of inquiry to three government agencies about the prospects of grant support for his research into obedience. His students built a fake "shock box" and he ran some preliminary studies using Yale undergraduates as subjects.

The money duly rolled in, and the experiment went ahead. Subjects were told that they were participating in a study of punishment and learning, in which they and another participant would play the role of either "teacher" or "learner," to be determined by drawing lots. What they did not know was that the other participant was a confederate of the experimenter, and that both pieces of paper said "teacher." The two participants were treated to a brief theoretical lecture about the role of punishment in learning and memory, then the putative "learner" was led away and the real subject of the experiment was seated in front of the shock box. This was an authentic-looking piece of equipment with a series of thirty switches labeled from 15 to 450 volts, grouped into batches of four switches with captions running from SLIGHT SHOCK to SEVERE SHOCK, the last two simply and ominously labeled xxx. The teacher was instructed to administer a simple word-association test and to punish the learner's wrong answers with shocks of increasing severity. The subjects were unable to see the learner, but his cries of anguish (prerecorded for consistency) were clearly audible. Famously, about a third of participants continued with the experiment up to the last set of switches.

Milgram claimed that his experiment was

> highly reminiscent of the issue that arose in connection with Hannah Arendt's . . . book *Eichmann in Jerusalem.* Arendt contended that the prosecution's effort to depict Eichmann as a sadistic monster was fundamentally wrong, that he came closer to being an uninspired bureaucrat who simply sat at his desk and did his job. . . . After witnessing hundreds of ordinary people submit to the authority in our own experiments, I must conclude that Arendt's conception of the *banality of evil* comes closer to the truth than one might dare imagine.[63]

He argued, that is, that he had exposed a universal psychological trait linking ordinary Connecticut folk to the defendant in Jerusalem: our tendency to obey those in authority.[64] The twist in the tale was that the role of Nazi-giving-orders was taken by a behaviorist psychologist. The real subject of the experiment, the hapless "teacher," was told, as he or she was seated at the shock box controls, that "psychologists have developed several theories to explain how people learn various types of material. . . . One theory is that people learn things correctly whenever they get punished for making a mistake."[65] The banality of evil in Milgram's memorable performance piece, in other words, was not bureaucratic European fascism but the routinized rituals of American pleasure-pain research.

The decade that began with Milgram's infamous experiment quickly generated other dramaturgical critiques of behaviorism. The following year, communist brainwashing was depicted in the film *The Manchurian Candidate* and aversive operant conditioning was portrayed in the novel *A Clockwork Orange*. In 1963 *Life* magazine ran an article on "Chemical Mind Changers," which opened with a riff on Frazier, the cranky, competitive founder of Skinner's fictional utopian community, Walden Two. Thereafter, Skinner's name was repeatedly invoked in relation to debates about behavior modification. He responded that so-called mind control happened everywhere, all the time, through the perfectly ordinary experiences of reward and punishment found in every environment. The only question that confronted the human species, he declared, was whether to design these environments scientifically or leave the whole business up to chance: "I merely want to improve the culture that controls."[66]

THE ROLE OF PLEASURE IN BEHAVIOR

The growing criticisms of behaviorism did not penetrate south to Tulane University, and Robert Heath continued with his research on human subjects undisturbed. In 1962 Heath organized a second symposium to showcase his work. Twenty-two scientists participated, including one member of the Montreal team that had accidentally discovered the rat's pleasure center a decade earlier. Heath presented his research in a paper entitled "Attempted Control of Operant Behavior in Man with Intracranial Self-Stimulation," which recounted his efforts to replicate the Canadian results in humans: "Thus far, reported self-stimulation work under controlled laboratory conditions has been confined to subhuman species. The present report describes exploratory efforts in the extension of such studies to man."[67] In July 1962 Heath's team had implanted depth electrodes in the brains of two human subjects, "a 35-year-old divorced white man with a diagnosis of chronic schizophrenic reaction, catatonic type," and "a 25-year-old single white man" suffering from "psychomotor epilepsy with possible underlying schizophrenia."[68]

The experiment went badly. To the frustration of the research team, "when the current was turned off, both subjects continued to press the lever at essentially the same rate." This was not what they were expecting. Why would the subjects press the lever when no reinforcement resulted from the action? Eventually, "the experiment was terminated by the experimenter after hundreds of unrewarded responses." When asked why they continued to press the lever, the "schizophrenic patient would invariably state that it felt 'good'; the epileptic subject would say that he was trying to cooperate with us, that he assumed we must want him to press it, since he had been placed there."[69] The scientists had invested their brightest hopes in the latter patient because he was "intellectually intact," but he turned out to be impossible to work with: "Besides frequent sullen and demanding behavior, he feigned adverse reactions to the stimulation (even on occasion with the current turned off)."[70] Most of the results were thus obtained from the other man, "B12," who presented problems of his own, as he "was catatonic, and might have sat for hours without making a movement that could be reinforced."[71]

Given the unanticipated challenge of getting the research subjects to stop pressing the lever after the current was turned off, the experimenters introduced a new protocol, which they dubbed, without apparent irony, the "free-choice procedure."[72] By this time they were working with only B12, the catatonic patient, whom they presented with two different things to press, a lever and a button, hoping that, allowed to choose, he might select the rewarding stimulation. This produced mixed results. When the choice was between rewarding and aversive current, B12 chose the positive stimulation, but when the choice was between rewarding current and no current, he evinced no preference. Heath and his team tried injecting B12 with methamphetamine hoping to enhance his sensitivity to positive stimulation, but "the drug produced no detectable change in response rate."[73]

The paper's disarmingly frank conclusion was that "attempts to establish, modify, and extinguish a simple lever-pressing response under conditions of intracranial self-stimulation in two human subjects have proved largely unsuccessful." Heath and his coauthors nonetheless asserted that "with revisions of procedure, data were obtained which suggest the presence of subcortical areas in the human brain in which brief electrical stimulation appears to have rewarding or reinforcing properties."[74] There was no discussion of the possible confounding effect of using human subjects, able to talk back, to resist, to obey, and, as on this occasion, to obey robotically as an exquisitely annoying form of resistance.

AUTONOMOUS MAN

On the East Coast, antibehaviorists congregated right under Skinner's nose at the Harvard Center for Cognitive Studies. In California they found a home near Stanford at the Center for Advanced Studies in the Behavioral Sciences in Palo Alto. Founded in 1954, the CASBS hoped to recruit fellows with all the qualities of open-mindedness that marked the antiauthoritarian personality: " the ability to handle anxieties of not knowing answers . . . to avoid premature closure, to tolerate areas of admitted ignorance."[75] For the 1964–1965 academic year they invited a man who, while not exactly excelling in the qualities of modesty, humility, and self-doubt, was certainly a great avatar of human freedom: the Hungarian-born novelist, journalist, and historian of science Arthur Koestler.

Koestler's engagement with the psychology of totalitarianism was deeply personal. He spent his twenties as a member of the German Communist Party. In the years of the great famine following Stalin's forced collectivization of agriculture, he traveled widely in the Soviet Union, where he "learned to classify automatically everything that shocked me as 'the heritage of the past' and everything I liked as 'the seeds of the future.' . . . All my friends had that automatic sorting machine in their heads."[76] In 1938, at the age of thirty-three, he finally abandoned the Communist Party, and with it the strictures of revolutionary greater-good reasoning.

His most famous novel, the 1941 *Darkness at Noon*, charts the political disillusionment of a middle-aged Bolshevik imprisoned by his former comrades:

> The "I" [was] a suspect quality. The Party did not recognize its existence. The definition of the individual was: a multitude of one million divided by one million. . . . It was a mistake in the system; perhaps it lay in the precept which until now he had held to be uncontestable, in whose name he had sacrificed others and was himself being sacrificed: in the precept, that the end justified the means. It was this sentence which had killed the great fraternity of the Revolution and made them all run amuck. What had he once written in his diary? "We have thrown overboard all conventions, our sole guiding principle is that of consequent logic; we are sailing without ethical ballast."[77]

Darkness at Noon was a critique of the Soviet version of "consequent logic." In 1967, after his stint at CASBS, Koestler published *The Ghost in the Machine*, his attempt to rebut the same totalitarian/utilitarian error in Cold War America. His target was the behavioral sciences, a "flat-earth view of the mind," which "has substituted for the erstwhile anthropomorphic view of the rat, a ratomorphic view of man."[78] Judging Skinner's educational philosophy "as nasty as it is naïve," Koestler proceeded to lay out an alternative view of man's place in nature. Drawing promiscuously from the work of any scientist who had said something convenient for its thesis, *The Ghost in the Machine* was a mixture of psychodynamic, evolutionary, and gestalt psychology, all served up with witty contempt for the author's opponents and eloquent enthusiasm for his allies. Koestler gave the last word to a pioneer of systems biology who had been a fellow at the CASBS: "What we need is not some hypothetical mechanisms better to explain some aberrations of the behavior of the laboratory rat; what we need is a new concept of man."[79]

The "concept of man" that stepped into the breach was not in fact "new" but centuries old: the vision of autonomy, dignity, and freedom to be found in the moral philosophy of Immanuel Kant. Kant's late-eighteenth-century concept of autonomy turned out to be a perfect fit with post-1945 anxieties about totalitarianism, especially with regard to medicine and psychiatry. The activists and reformers of the 1960s found that their disgust with the utilitarian calculus of human experimentation resonated precisely with the philosopher's injunction that no rational being should be used as a means to an end. Kantianism seemed perfectly to encapsulate the spirit of informed consent and the sanctity of autonomous decision-making.

Indeed, anyone interested in contemporary medical ethics might be struck by how much of Kant's *Groundwork of the Metaphysics of Morals* seems applicable to the principle of informed consent. The essay opens with an anti-utilitarian thought experiment about the existence of pure good will, arguing that the concept of goodness is independent of all consequences: "A good will is not good because of what it effects or accomplishes, because of its fitness to attain some proposed end, but . . . it is good in itself."[80] Having persuaded the reader that there is a such a thing as a pure good will, independent of consequences, Kant builds up a picture of virtue as a rational system of duties—some absolute, some optional—to oneself and to others. As a counterpart to his analysis of obligations, he analyzes rights, especially the absolute right to "not be used merely as a means."[81] On a practical level this entails doing unto another rational being only those things that she can agree to, in the fullness of her freedom.[82]

Kantian autonomy resonated even more strongly in radical psychology than in reformist medical ethics. In 1965 psychologist Thomas Szasz published *The Myth of Mental Illness*, a tract on the rights of autonomous man and the virtues of the Kantian therapist. "Autonomy," he announced, "is a positive concept. It is freedom to develop oneself—to increase one's knowledge, to improve one's skill, and achieve responsibility for one's conduct. And it is freedom to lead one's own life, to choose among alternative courses of action as long as no injury to others results." Szasz even borrowed from Kant the technical term "heteronomy" as autonomy's opposite. In Kant's formulation, heteronomy was the moral error of acting according to rules flowing from an external source. To behave virtuously in order to store up riches in heaven, for example, is to act for a heteronomous reason. Most importantly in relation to the human experimentation scandals, acting based on the calculation of consequences— the essence of utilitarianism—is just as bad as acting in obedience to any other external reward. In Kant, anti-utilitarian psychology and medicine found its philosophical north star.[83]

BEYOND FREEDOM AND DIGNITY

Despite the swelling ranks of his critics within academia, the 1960s saw steadily mounting enthusiasm for Skinner's work outside the ivory tower. Sales of *Walden Two* began to rise, and by the middle of the decade, he had taken on the unlikely role of hippie prophet. By decade's end a series of intentional communities had been founded, explicitly inspired by the communitarian ideals of the novel. Bright hopes for behaviorist communes dimmed as fast as any, however. Most ran into trouble when the communards discovered that Skinner's antidemocratic approach—in which a small cadre of planners and managers discreetly ran the show— was unworkable in the context of American 1960s counterculture. It turned out that anyone predisposed to start or join such a community

either thought that he or she should be the Frazier figure or, more mod-
estly, desired to share equally in the decision-making.[84]

The antidemocratic core of Skinner's vision might have been a practical failure, but on college campuses his appeal to the young only increased
over the final, apocalyptic years of the sixties. Around 1970 his lectures
moved from seminar rooms to football stadiums. In 1971 he published a
summary of behaviorist philosophy that stayed on the *New York Times*
best-seller list for twenty weeks. The provocatively titled *Beyond Freedom
and Dignity* laid out his utilitarian blueprint for solving the "terrifying
problems that face us in the world today," including population explosion, nuclear holocaust, famine, disease, and pollution. The humanists,
he scoffed, appealed to explanations such as "attitudes," "opinions," and
"personalities" to account for self-destructiveness. Skinner's aim was
to replace the misleading language of internal mental states with un-
adorned descriptions of human behaviors. That year he made forty radio
and television appearances.[85]

Skinner's principal target was a specter he called "autonomous man,"
which he derided as an Enlightenment superstition, heaping contempt
on "the literature of freedom."[86] "Autonomous man," he sneered, "serves
to explain only the things we are not yet able to explain in other ways."[87]
"Responsibility" was another symptom of our pathetic ignorance. If we
knew how criminality was caused, we would not need any such vacuous
notion. "We shall not solve the problems of alcoholism and juvenile delinquency by increasing a sense of responsibility. It is the environment
which is 'responsible' for the objectionable behavior."[88] The metaphysical
illusion of moral freedom was not only irrational, it was pernicious, im-
peding "the design of better environments rather than of better men."[89]

One of the most ticklish questions Skinner confronted in the book was
how to justify his prescriptions in terms that he deemed appropriate to
science. Scientific credibility required that he present his plans for a better
world as somehow value-free. The title *Beyond Freedom and Dignity* was
a tribute to Friedrich Nietzsche's brilliant manifesto of amorality *Beyond
Good and Evil*, but Skinner spake more like Hobbes than Zarathustra.
His argument was completely Hobbesian: goodness was just a word for
pleasantness, and the whole thing was predicated on brute survival. He
concluded that his behaviorist prescriptions concerning pollution, pop-

ulation, and nuclear war were not a vision of the good life but merely an expression of his evolutionarily programmed and culturally reinforced desire for the survival of his species and of his culture.[90]

Although Skinner was as utopian as Joseph Priestley, as messianic as Jeremy Bentham, and as convinced of the perfectibility of man as Herbert Spencer, his determination to strip even human happiness of moral content left his followers bereft of substantive reasons to enact his program of social reform. His supporters outside the academy wanted desperately to build a better world, but their scientism proved unequal to the moral eloquence of the avatars of freedom, dignity, and autonomy joining forces against them.

THE INITIATION OF HETEROSEXUAL BEHAVIOR

Meanwhile, Heath was still doggedly working out the therapeutic possibilities for direct electrical stimulation of the brain. In 1970 he embarked on a study to change a young man's sexual orientation, employing "pleasure-yielding septal stimulation as a treatment modality for facilitating the initiation, development, and demonstration of . . . heterosexual behavior in a fixed, overt, homosexual male."[91] This work betrayed the lasting influence of Heath's mentor Sandor Rado, who took a highly paternalistic and judgmental approach to sexuality.

Rado was perfectly permissive when it came to heterosexual pleasure. He judged the clitoral orgasm to be "just as respectable and desirable as any other kind of female orgasm." He also opined, unappetizingly, that "it is hard to see from a medical point of view just why the mouth-genital contact should be more objectionable than the contact of two bacteria-filled mouths." When it came to same-sex desire, however, he expressed nothing but scorn. In attempting to update Freud, Rado had hitched his theory of pleasure to such a narrow concept of evolutionary utility that the idea of nonprocreative desire was unthinkable: "I know of nothing that indicates that there is any such thing as innate orgiastic desire for a partner of the same sex."

For Rado, homosexuality was a pathology akin to schizophrenia, complete with its own compensatory delusions of grandeur: "The homosexual male often clings to the myth that he belongs to a third sex, superior to the rest of mankind. This would seem to be the effort of an individual who lives in constant dread of detection and punishment, which is the milieu of the society that prohibits homogenous mating, to restore his shaky equilibrium."[92] Partly under the influence of Rado's evolutionary theory of sexuality, reparative therapy enjoyed a vogue in the mid-twentieth century, with psychiatrists prescribing talk therapy, hormones, drugs, and other treatments to "restore" their patients' heterosexuality. In this context, electrical brain stimulation seemed to promise a new horizon of efficacy.

To test the reparative powers of his techniques, Heath had "patient B19" put under general anesthetic while electrodes were implanted in eight deep brain structures. Four weeks later, once the surgery had healed but before the electrodes were connected to any current, he was made to watch a fifteen-minute "'stag' film, featuring sexual intercourse and related activities between a male and female." At the end of the experience "he was highly resentful, angry, and unwilling to respond." The following day he embarked on a schedule of stimulation of the septal region, sometimes administered by the researchers, sometimes self-administered. He "exhibited improved mood, smiled frequently, stated that he could think more clearly, and reported a sense of generalized muscle relaxation. He likened these responses to the pleasurable states he had sought and experienced through the use of amphetamines." Eventually, "he had to be disconnected, despite his vigorous protests." Most importantly, he "reported increasing interest in female personnel and feelings of sexual arousal with a compulsion to masturbate," after which "he agreed without reluctance to re-view the stag film, and during its showing became sexually aroused, had an erection, and masturbated to orgasm."[93]

Over the next few days B19 reported "continued growing interest in women." After another series of septal stimulations, he expressed "a desire to attempt heterosexual activity in the near future." The researchers accordingly arranged for a twenty-one-year-old prostitute to have sex with him in a laboratory while he was hooked up to an EEG apparatus to measure his brain activity. She seems to have been most therapeutically adept, allowing him to talk to her for an hour about "his experiences with drugs, his homosexuality and his personal shortcomings and negative qualities," during which she was "accepting and reassuring." In the second hour, "in a patient and supportive manner, she encouraged him to spend some time in a manual exploration and examination of her body." They had sex and, "despite the milieu and the encumbrance of the electrode wires, he ejaculated."[94] Heath declared the experiment a success: "Of central interest in the case of B19 was the effectiveness of pleasurable stimulation in the development of new and more adaptive sexual behavior."[95]

For a protégé of Rado, it made sense to lump in homosexuality with depression, paranoia, grandiosity, and suicidal tendencies, as just another evolutionarily maladaptive disorder of hedonic regulation. As the study was under way, however, a campaign on the part of gay activists succeeded in getting homosexuality removed from the Diagnostic and Statistical Manual, the so-called Bible of psychiatry. The DSM-III, published in 1973, contained a new diagnosis of "Sexual Orientation Disturbance (SOD)," suffered by "individuals whose sexual interests are directed primarily toward people of the same sex and who are either disturbed by, in conflict with, or wish to change, their sexual orientation."[96] By pivoting on lack of self-acceptance, this diagnostic category implied that treatment for SOD might consist of a few affirmative sessions with a gay therapist rather than any sort of procedure aimed at changing sexual orientation.[97] SOD was perfectly symptomatic of the state of medical ethics in 1973, replacing reparative therapy with an ethic of choice and self-determination.

ZERO PLUS ZERO EQUALS ZERO

Noam Chomsky's review of Skinner's 1971 best seller, *Beyond Freedom and Dignity*, made his earlier review of *Verbal Behavior* look like the model of academic politeness. Chomsky demanded rhetorically what the book contributed to psychology, and then answered his own question with a triple negative: "zero plus zero still equals zero." As for Skinner's proposals for societal reform, he summarized them as "a well-run concentration camp with inmates spying on one another and the gas ovens smoking in the distance." He admitted that "it would be improper to conclude that Skinner is advocating concentration camps and totalitarian rule" but only because this "conclusion overlooks a fundamental property of Skinner's science, namely, its vacuity."[98]

It was despair and anger over the Vietnam War that had sent Chomsky and his fellow travelers to such rhetorical extremes, dividing the world into those who opposed the fascism of American foreign policy and those who collaborated with it. By denying that true political resistance was even possible, behaviorists were automatically counted as collaborators with oppressive conditions at home and abroad. Chomsky's damning review was a nail in the coffin: 1971 was the year that witnessed the first steep decline in articles and PhD dissertations on behaviorist psychology, mirrored by a corresponding rise of the cognitive school.[99]

Protesters began to attend Skinner's public lectures, including a splinter group of the radical left-wing group Students for a Democratic Society. The National Caucus of Labor Committees delivered a "Nuremberg Indictment for Crimes against Humanity" against him. On the campus of Indiana University, activists hanged him in effigy. At one talk, a protester fired a blank cartridge during a question and answer session. Police began to stand by at Skinner's public appearances, and he made sure to walk to his office at Harvard by a different route every day.[100] In 1974 a defense of his philosophy entitled *On Behaviorism* appeared. Radical psychologist Thomas Szasz, the champion of neo-Kantian therapy (discussed above), wrote in the *Libertarian Review*, "I believe that those who rob people of the meaning and significance they have given their lives kill them and should be considered murderers, at least metaphorically. B. F. Skinner is such a murderer. Like all such murderers, he fascinates, especially his victims."[101]

In January 1973 the prestigious journal *Science* published an account of an experiment that did for psychiatry what the Tuskegee exposé had done for medicine. The report was entitled "On Being Sane in Insane Places" and described how eight "pseudopatients" had made appointments at a total of twelve psychiatric hospitals. Upon arrival at the admissions office, the ersatz patient complained "that he had been hearing voices. Asked what the voices said, he replied that they were often unclear, but as far as he could tell they said 'empty,' 'hollow,' and 'thud.'" The choice of these words was "occasioned by their apparent similarity to existential symptoms."

Apart from reporting this one auditory hallucination, the pseudopatients then proceeded to do nothing that they would not normally do. The details of their lives to date were recounted as faithfully as possible; they cooperated with staff; they were friendly to the other patients. Asked about their symptoms, they said that the voices had gone away. All but one were admitted with a diagnosis of schizophrenia and discharged, after an average of nineteen days, with a diagnosis of schizophrenia in remission. None were detected as frauds by the doctors, although thirty-five fellow patients guessed that the participants were journalists or academics engaged in some kind of muckraking investigation. The psychiatric establishment was revealed as not just cruel and controlling but also chronically obtuse.[102]

The net was closing. The same month as the publication of "On Being Sane in Insane Places," CIA director Richard Helms destroyed the records of project MKULTRA, the agency's covert human experimentation in the 1950s and 1960s on psychotropic drugs and brainwashing techniques.[103] Meanwhile, down in New Orleans, Robert Heath was finally beginning to experience the kind of treatment dished out to his colleagues in the north. An organization called the Medical Committee for Human Rights (founded in 1964 to provide treatment for wounded civil rights activists) came across the reports of Heath's experiments in a scientific journal and began to follow his work. In May 1972 Heath was scheduled to speak at a conference at a French Quarter hotel. The activists circulated a letter with some details of the sexual orientation study, claiming that the research subject had been busted for drugs and had agreed to enter Charity Hospital for psychiatric observation only in order to rid himself of the charges.[104] They organized a demonstration at the conference; Heath kept away. Tulane University was not available for comment.

A local journalist subsequently published a long account of Heath's work, characterizing him as "a publicity-shy psychiatrist," a ghoulish and secretive figure floating through Charity's mental wards in search of experimental subjects.[105] The focus of much of the activists' outrage was the research review process at Tulane, which left "innocent patients and inmates helpless and open to abuse." The "release of claims" document that Heath's patients had to sign, they argued, gave him "carte blanche permission for implantation, surgery, drugs and other treatments."[106] Far from protecting the patients, the research review process seemed designed to give Heath legal cover.

The article was entitled "The Mysterious Experiments of Dr Heath: In Which We Wonder Who Is Crazy and Who Is Sane," echoing "On Being Sane in Insane Places." This inversion of madness and reason was the baffled howl of a generation born in the shadow of the Holocaust, who grew up being bombarded with denunciatory rhetoric about communism, and then discovered in young adulthood that capitalist democracy had enacted its own version of totalitarian rule.

In its conjoined cynicism, utopianism, and wild innocence, the antipsychiatry movement was the distilled essence of the 1960s counterculture as it came to an explosive finale in the Watergate era. For those of an activist bent, paternalistic medicine seemed little more than a thin veneer for the exercise of patriarchal control. The deceived and coerced black biomedical research subject became the emblematic victim of American prejudice and hypocrisy. Heath's young, gay, white victim took his place alongside the Tuskegee subjects, as a representative of the oppressed and silenced masses now rising up against the forces of social control. The demand for informed consent in research had come together with the movement against coercive psychiatry in a denunciation of the whole medical, psychological, and political establishment.

In 1975 the American Civil Liberties Union published a full-length book entitled *The Rights of Hospital Patients*. After "some uncontrollable or unexpected event," the book explained, the patient is admitted to a hospital not of his choosing, made to sign a variety of forms without explanation, separated from friends and relatives, placed in a room, stripped of clothes, made to wear a one-piece garment "designed for the convenience of hospital staff," confined to a bed, made to ask permission to use the toilet, and "given a wrist band with a number written on it, a number that may become more important than the patient's name."[107] According to this melancholy view of medical authoritarianism, the ordinary hospital patient in 1970s America faced a predicament almost as profound and as troubling as that of the embattled hero of Arthur Koestler's *Darkness at Noon*.

The same year, the film *One Flew over the Cuckoo's Nest* appeared. Based on a 1962 novel by Ken Kesey, it chronicles the liberation of a group of patients in a psychiatric ward, jerked out of their apathy by the hell-raiser Randle "Mac" McMurphy, played by Jack Nicholson. The arc of the story concerns McMurphy's battle of wills with the icy ward sister, but its emotional center is Chief Bromden, a Native American inmate. Bromden—played by the actor, artist, and rodeo performer William Sampson—is a man of majestic appearance, approaching seven feet tall, whom everyone believes to be deaf and dumb. He is known as "Chief

Broom" because he spends his days pushing a broom wrapped in a white cloth slowly around the ward. McMurphy teaches him to shoot hoops, and—in one of the best feel-good scenes in the history of cinema—the inmates win a basketball game against the African-American ward aides.

One day, as they sit on a bench together, McMurphy jostles Bromden's elbow in order to offer him a stick of gum. Bromden takes the gum, and quietly thanks McMurphy. McMurphy starts with astonishment, offers the Chief another piece of gum, and is rewarded with Bromden looking down and murmuring in a honeyed growl, "Yeah . . . Juicy Fruit." "You sly son-of-a-bitch, Chief! Can you hear me, too?" McMurphy cries. "You bet," Bromden answers, looking the other man in the eye. McMurphy turns his head away, whoops with joy, slaps his thigh, looks back at Bromden, and says, "Oh my *god damn*, Chief! . . . You fooled them, Chief. You *fooled* them, you fooled them *all*!" The dumb Indian is revealed to be the smartest man in the room, and McMurphy promptly starts planning to escape with him to Canada.

Kesey wrote the novel at the age of twenty-three after a stint working a graveyard shift at the Veteran's Hospital in Menlo Park, California. It was 1961: the age of the Eichmann Trial, of Milgram's "obedience to authority" experiments, and of Heath's symposium at Tulane on "the role of pleasure in behavior." The book is narrated by Bromden and mounts an argument about psychiatric totalitarianism in narrative form. Its indisputable prophetic power is marred only by its casual racism toward the aides, the "Black Boys," whom Kesey portrays as sadistic lackeys.

By the time the film was made, the antipsychiatry movement had joined forces with the civil rights movement, and the black aides had perforce become more sympathetic characters, portrayed as cool and cynical in black bow ties and crisp white shirts. In the decade and a half since the book's publication, opposition to the Vietnam War had drastically simplified the moral universe. On one side stood Randle McMurphy and his fellow travelers, wild spirits of youth, justice, and freedom; on the other stood the Nurse Ratcheds of this world, America's Cold War version of fascist rule. Chief Broom—the mute indigenous hero who had witnessed the American jackboot crushing his family and his people— was the perfect symbol of the sense of universal solidarity that underlay the movement for informed consent. And when he finally spoke, it was as if the silenced peoples of the world spoke with him.

6

Epicurus Unchained

THE WHITEWASH

In September 1973 the National Research Act, mandating informed consent, passed the Senate. The Kantian individual—free, rational, possessed of inalienable dignity, never to be used as a means to an end, always to be treated as an end in herself—had triumphed. Of course, this meant she had to be tamed. Radical solidarity with the wretched of the earth was all very well for kooks like Kennedy, but there was serious regulatory business to be done, and the practice of informed consent had to be rendered usable by the notoriously conservative American medical profession.

A committee was duly appointed, charged with getting the issue of human experimentation "off the floor of the Congress."[1] For the first two years of its existence, the commission met in subcommittees on single issues such as experimentation on fetuses and research in prisons. In 1976 the whole membership convened at the Belmont Conference Center in Maryland for an intensive four-day retreat, in order to hammer out the basic principles on which medical ethics should be founded. They came up with three: "respect for persons," "beneficence," and "justice." After the retreat, they engaged the services of an academic philosopher, Thomas Beauchamp, a specialist in the work of David Hume, to provide an intellectual justification for the three principles. As befits a Hume scholar, Beauchamp was an avowed utilitarian.

Hard to say why the commission picked a utilitarian for the job. It was surely an odd choice, given that the reforms were necessitated by the abuse of greater-good reasoning, but probably it was just philosophical obtuseness on their part. Whatever the reason he was appointed, and in defiance of his own philosophical convictions, Beauchamp did not recommend a utilitarian framework for the new medical ethics. In the "Belmont Report"—named for the conference center—he defined the first principle of medical ethics as "the requirement to acknowledge autonomy and the requirement to protect those with diminished autonomy." The application of this principle, of course, was informed consent. Thirty turbulent years after the publication of the Nuremberg Code, American medicine had finally caught up with its own professed values.

Beauchamp was forced by circumstances to put autonomy first, but "beneficence"—in the form of a cost-benefit analysis of the risks and rewards of research—made it to a close second place. The anti-utilitarian thrust of the human experimentation scandals necessitated this ranking of the principles—autonomy first, utility second—but the Belmont Report was crucially vague as to *why* the rules of medical conduct should be ordered that way, leaving utilitarianism bloodied but unbowed. As for the historical events leading up to the reforms, Beauchamp mentioned the Tuskegee syphilis study as something that had taken place "in the 1940's," a face-saving pruning back of the true chronology of 1932–1972.[2]

At the same time as he was composing the Belmont Report, Beauchamp was teaching courses in bioethics with the Quaker theologian James Childress. The two men found that they "shared a passion" for teasing out the ethical theories behind any given position or argument, and they began to collaborate on a book.[3] Childress was the author of the 1970 paper entitled "Who Shall Live When Not All Can Live," a response to a Seattle hospital's deliberations about which desperately ill kidney patients should receive dialysis treatment. As discussed in the first chapter, the "God Committee" had appealed to social-worth criteria in making its decisions, and became an unfortunate punching bag for the patients' rights movement. Childress's paper, arguing for random selection instead of judgments of worthiness, was a classic bit of anti-utilitarian reasoning, in which his Quaker ideal of the sanctity of the human spirit was translated into secular, Kantian terms. "God might be a utilitarian," Childress had declared, "but we cannot be."

Childress's collaboration with Beauchamp thus brought a utilitarian and an anti-utilitarian together to figure out the new medical ethics. Their coauthored book, which came out in 1979 under the title *Principles of Biomedical Ethics*, was a work of delicate compromise. "So far as our actions in regard to others are concerned," the authors observed, "it is doubtful that the approaches taken by Mill and Kant lead to significantly different courses of action."[4] Despite this claim of moral equivalence, the ranking of the principles remained the same, with respect for autonomy and informed consent topping the bill. When it came to the second principle, the book deviated slightly from the Belmont Report, dividing it into two—"nonmaleficence" and "beneficence"—on the grounds that not hurting people is more important than helping them. The fourth principle, that of "justice"—always a bit of a redheaded stepchild—was defined in a bunch of different ways and ended up somewhere between diffuse and meaningless. So in theory there were four principles, but in practice all the substantive discussion in the book was about the tensions between utility and autonomy, mostly discussed in terms of the limits on paternalism.

Beauchamp and Childress may have denied there was any contradiction between Mill and Kant, but they did eventually get around to defining the duty to obtain informed consent in anti-utilitarian terms:

> There is a moral duty to seek a valid consent *because* the consenting party is an autonomous person, with all the entitlements that status confers. By contrast, neither utility nor nonmaleficence leads to this strong conclusion, for both would justify not seeking consent in some circumstances— utility when it would not maximize the social welfare, and nonmaleficence when no apparent harm would result.[5]

This was the soft-Kantian move that gently, almost inaudibly, placed autonomy at the head of the four principles. And therein lay the pragmatic wisdom of the *Principles of Biomedical Ethics*. It enshrined the requirement to seek informed consent as the threshold condition for any kind of medical action while making peace with the long utilitarian history of medical research. The book successfully rehabilitated cost-benefit thinking, but patient autonomy was perforce at the heart of its approach.

The triumph of patient autonomy has been resoundingly confirmed in everyday practice. Informed consent grows ever stronger, wider, and more rigorous, far outstripping Beauchamp and Childress's original conception. The first edition of their book fell quickly out of step with changing ethical standards. One section, for example, valorized the hepatitis research on children at the Willowbrook School. The story is well known. Between 1955 and 1970, New York University researchers gave live hepatitis virus mixed into chocolate milk to children who were wards of the state at a school for the "mentally retarded" on Staten Island. The researchers maintained that conditions at the school were so appalling that the children were destined to become infected anyway (a claim belied by their own estimates of a background infection rate at the school of 25 per 1,000).[6] The research was productive, leading to the discovery of the two strains of the virus now known as hepatitis A and B. About this episode (one of Henry Beecher's examples of unethical research in his 1965 whistle-blowing piece), Beauchamp and Childress commented "it is generally agreed that both individual and concerted social action intended to benefit others are sometimes morally justified, even though they involve certain risks."[7] Over the next few years, as informed consent protocols got progressively stricter, the Willowbrook episode decisively entered the annals of human experimentation abuse, and later editions of the book omitted this affirmative language.

In keeping with its peace-making agenda, the book was even more tactful than the Belmont Report about the scandal-ridden history that gave rise to its own production, summing it up in two sentences: "Since the Nuremberg Code of 1947, the United States Department of Health, Education, and Welfare has promulgated several guidelines for research involving human subjects. In 1974 Congress created the National Commission for the Protection of Human Subjects of Biomedical and Behavioral Research."[8] Informed consent, the authors announced, arose directly from "the horrible story of experimentation in concentration camps." In the United States, they claimed, it "has gradually emerged from malpractice cases."[9] The most important case referred to was a suit brought against Stanford University in 1957 by a patient whose surgery had gone wrong. Tuskegee was not even mentioned. By juxtaposing Nazi experimentation with American malpractice cases—contrasting a callous Old World with an almost oversensitive medical profession on the domestic front—the authors drew a discreet veil over the history of human experimentation in the United States. Tuskegee finally limped into the third edition of 1989, in a very brief aside.

Don't get me wrong. I am a grateful beneficiary of the more respectful medical treatment that has resulted from the application of the four-principles approach. But I will say this: by evading the history that made reform necessary, the authors failed to acknowledge that the principle of autonomy is not just a polite add-on to medical utilitarianism but a substantive refutation of its moral supremacy. We cannot do without informed consent. We also cannot do without medical utilitarianism. Beauchamp and Childress were right about the need for a scheme that incorporated both principles. But the reasons for autonomy's primacy have been willfully obscured in bioethics by a historical narrative sanitized to the point of falsehood.

If the Belmont Report whitewash stands at one pole of storytelling, at the opposite pole stand those historians who continue to denounce Cold War American medicine as truly fascistic. One side has painted a picture of a benign American medical profession responding to individual cases of clinical malpractice with sensitive and far-reaching legal reforms. The other side cannot discern much salient moral distinction between, say, cancer researcher Chester Southam and Hitler's physician Karl Brandt. This is, of course, a bit of a caricature. Neither side is quite that extreme. It is, however, no exaggeration to say that the problem with utilitarianism slips out of the grasp of both sides like a greased pig. And now, four decades after the Belmont Report, the pig has come sauntering back—smelling like a rose.

UNLEASHING PHARMACEUTICAL POWER

Hot on the heels of the Belmont Report, patient autonomy began to be redefined in economic terms. This was inevitable given the dramatic developments in the Cold War: mandatory informed consent shares an exact horoscope with the global transformation that goes by the name of "neoliberalism." On September 11, 1973, the same day the National

Research Act passed the Senate, a CIA-backed coup overthrew the socialist government of Chile and replaced it with a military dictatorship hell-bent on privatizing the economy. A month later, in October, came the Arab-Israeli war, the subsequent oil embargo, and the fuel crisis. After the oil crash, the social-democratic version of capitalism that had prevailed since the last years of the Great Depression was dismantled in favor of a new brand of free-market fundamentalism. Between 1978 and 1980 China, the United Kingdom, and the United States began to deregulate their economies and to sell off public assets. In 1990 the Soviet Union collapsed. By the end of the century, in harmony with the global victory of laissez-faire economics, the sovereign individual at the center of reformist medical ethics had quietly metamorphosed from informed citizen to empowered consumer.

Much of the transformation took place in relation to AIDS activism, a story replete with instructive reversals and ironies. *Principles of Biomedical Ethics* was first published in 1979. In July 1981 the Centers for Disease Control identified a new disease so deadly that it would turn the idea of patient autonomy as the right to *refuse* treatment upside down. The mystery surrounding AIDS made its lethality all the more terrifying. At first, all that scientists had to go on in their search for a cause was the roughest kind of epidemiology. *Who* succumbed to this condition that systematically dismantled the immune system? *Where* were these rare cancers and pneumonias showing up? The answer seemed to be that gay men living in the urban villages of New York and San Francisco were uniquely susceptible to the disease.

Scientific bafflement, combined with the pathologization of an already despised minority, made the appearance of "gay related immune disorder" (as AIDS was then known) ripe for the revival of the defiant medical activism of the early 1970s. Even when the virus was identified in 1984, the years of a test but no cure only fostered more therapeutic pessimism. Resolute governmental silence plus vocal fear and prejudice on the part of some members of the medical profession did nothing to allay the paranoia of the gay community. The same year as the identification of the viral cause of AIDS, for example, public health officials seeking to close San Francisco's gay bathhouses were accused of plotting a homophobic conspiracy.[10]

The glittering career of fashion model turned New Age guru Louise

L. Hay is a measure of the depths of the antiscience sentiment provoked by AIDS. An ordained minister of the Church of Religious Science, Hay claimed that she had healed herself of vaginal cancer by releasing the resentment she carried into adulthood after having been sexually abused as a child. Her message was that AIDS, too, could be healed with love and forgiveness. Sometime in the middle of the 1980s, she started to meet with a group of gay men in her living room in Santa Monica, California, to practice affirmations and visualizations. The meetings ballooned, eventually turning into weekly "Hayrides," drawing hundreds of people to a West Hollywood auditorium every Wednesday evening.

A radiant sixty-something dressed in flowing rainbow robes, Hay would seat herself on a pillow and outline her self-love prescriptions. Prayers were said for the dying, and audience members delivered testimonials celebrating normal T-cell counts or reversions of their antibody tests to negative. At the end of the evening, the chairs were stacked against the walls and the participants arranged themselves on the floor in concentric circles, intoning "Om" to the center of the circle, where the most urgently afflicted members of the fellowship were lying.[11]

Hay was the high priestess of a cult of pure magical thinking. Her followers tended to be more eclectic, combining her spiritual techniques with a desperate search for effective drug treatments. People with AIDS and their supporters started grass-roots organizations such as Project Inform, which disseminated highly technical scientific articles among their members, tracking new black markets in experimental drugs. One promising drug was available in Mexico, in the borderlands *farmacias*. Another circulated among the gay community in Paris. The movement for patient autonomy had begun as a right to refuse the coercions and blandishments of medicine; now these activists desperately wanted in, insisting that "patients interested in trying experimental drugs should have the right to assume risks rather than endure the benevolent protection of the authorities."[12]

In a revealing twist on the rhetoric about Nazi human experimentation over the previous two decades, rabble-rousing gay activist and playwright Larry Kramer announced that genocide could be achieved by sins of omission as well as commission: "a holocaust does not require a Hitler to be effective. . . . Holocausts can occur, and probably most often do occur, because of *inaction*. This inaction can be unintentional or deliberate."[13] For most of the twentieth century, the American medical profession had been an active participant in nonconsensual human experimentation. At the height of the movement for informed consent, these violations of the Nuremberg Code earned them comparisons with the Nazis. Now, in the midst of a terrifying epidemic that wasted its victims to the appearance of death-camp inmates, it was the perceived *lack* of action on the part of medical researchers that made them seem fascistically callous all over again.

In May 1986 a newsletter called *AIDS Treatment News*, edited by a former computer programmer in San Francisco, noted that there was already a "effective, inexpensive and probably safe" drug against the disease, an unsuccessful anticancer agent called AZT that had been taken down from a shelf at the pharmaceutical giant Burroughs Wellcome and shown to have some action against HIV in vitro. But FDA regulations, the article claimed, meant that it would take at least two years for the drug to be licensed, during which time tens of thousands of preventable deaths would have occurred. The announcement was followed by a call for people with AIDS to take their place at the table alongside scientists and bureaucrats: "The companies who want their profits, the bureaucrats who want their turf, and the doctors who want to avoid making waves have all been at the table. The persons with AIDS who want their lives must be there, too."[14]

Since the thalidomide disaster, FDA regulations for getting new drugs to market had, indeed, become extremely burdensome. After extensive animal testing, trials for safety would be undertaken on a few human subjects. Placebo-controlled Phase 2 trials for safety and efficacy would then be done on hundreds of patients, with neither doctor nor patient knowing who was getting the placebo and who the real drug. Only if the drug looked promising at this stage would Phase 3 kick in, consisting of large-scale trials on thousands of patients. By the end of this process, the data submitted to the FDA could run into dozens of thick volumes, which could take months or years to review.

AZT went a different way. A few months into the Phase 2 trials, nineteen of the patients on the placebo had died against only one on AZT. On September 19, 1986, the blinds were stripped from the study and everyone on the placebo received the drug. The requirement for Phase 3 trials was waived, and in March 1987 AZT was licensed, bringing the process down from eight to two years.[15] The accelerated approval process created a long legacy of uncertainty. AZT had terrible side effects, including suppressing bone marrow, causing severe anemia. Most AIDS patients were desperate to get hold of the drug, but there was a vocal minority of refuseniks, who regarded it as iatrogenic poison and as only more evidence of the brutal indifference of the scientific establishment.

The same month as the release of the drug, a militant new patient rights group, the AIDS Coalition to Unleash Power, or ACT UP, met for the first time at the Gay and Lesbian Center on 13th Street in downtown Manhattan. The New York counterpart of the California Hayrides, ACT UP was kitted out in black leather rather than rainbow chiffon and preached militant direct action instead of personal self-forgiveness. I attended some of those early meetings and can attest to the extraordinary sense of energy and purpose in the room. The organization's first demonstration was on Wall Street, where protesters rallied to demand "immediate release by the federal Food and Drug Administration of drugs that might help save our lives."[16]

Alienated young revolutionaries in the AIDS activist movement burrowed through mountains of technical literature, a more striking achievement in the days before the Internet. Physicians and researchers learned to take these punk queers seriously, finding that they were often better informed than the certified experts. The activists and the medical professionals began to collaborate on community-based drug trials. They worked out a system whereby trial participants together with their primary care doctors set their own policies about placebos. In theory this resulted in a loss of scientific rigor; in practice it ensured greater levels of compliance on the part of the research subjects.

This was just the opportunity that advocates for pharmaceutical deregulation had been waiting for. Since 1972 free-marketeers at the University of Chicago had been agitating for the dismantling of FDA regulations. The thrust of their neoliberal argument was that the market was always a more efficient system of information dissemination than anything a government might put in place, and that by keeping valuable drugs from being licensed the FDA was causing more illness and death than it prevented. In 1973 they published a volume of conference proceedings making the case for comprehensive deregulation of the drug approval process, which was completely panned in the medical press.[17]

Figure 14. ACT UP protest at the FDA office in Los Angeles, January 29, 1988. The photograph captures some of the urgency and fury that transformed medical ethics in the 1980s. The thrust of the 1970s reforms had been to protect ordinary citizens against being enrolled in experiments and trials. In the context of the AIDS epidemic, protectionist regulations started to look like lethal bureaucratic obstructionism. Comparisons with Nazi medicine reappeared, but this time it was lack of action that made the medical profession seem fascistically callous. The irony of the AIDS movement's antipaternalism was that activists who fought against profiteering in the pharmaceutical industry helped push through a deregulatory agenda that ultimately served the industry bottom line. Photo by Chuck Stallard, courtesy of ONE Archives at the USC Libraries.

The economists were so out of step with the times that they were dismissed as irrelevant.

Almost twenty years later, AIDS changed all that, as desperately ill activists, motivated by nothing less than the desire to stay alive, engaged in a critique of the FDA and its paternalistic protections that was in all practical respects identical to the neoliberal position.[18] In 1992 the US Department of Health and Human Services announced initiatives "to provide earlier access to important new drugs, ease unnecessary regulatory burdens and strengthen US competitiveness."[19] As one of the FDA commissioners at the time later observed, "So you had this synergy between these ultra-liberal AIDS activists and these almost right-wing conservatives who wanted regulation reduced."[20]

In 1992 the FDA announced that ddC, an antiviral drug developed by Hoffman-La Roche, had been approved under licensing procedures that had been reduced from about two years to a mere eight months. The process was speeded up through the use of "surrogate endpoints" as evidence of efficacy.[21] Surrogate endpoints are a gift to the pharmaceutical industry. Instead of having to prove better clinical outcomes with the use of a given drug—longer life, better health—all a company has to do is to show improved in-vitro test results. In the case of AIDS, the relevant test might be the amount of detectable virus or number of T-cells in the blood. Not only is this procedure faster, it is also cheaper and more flexible, leaving more discretion in the hands of industry executives. In the context of the AIDS epidemic, the strategy helped save lives. A quarter century later, it also saves a great deal of drug company money.[22]

In the mid-1990s, the lead scientist at a private research foundation in New York City began to champion the idea of attacking the virus at different phases of its life cycle. With the success of this treatment protocol—the "triple cocktail"—AIDS ceased to be a death sentence. The urgency and fury that had fueled AIDS activism inevitably cooled, but its libertarian twist on medical antipaternalism left an indelible imprint. In 1973 the autonomous patient was a free-speech-movement rebel, saying "no!" to an all-powerful scientific establishment. By 1987, in the midst of the epidemic, his overwhelming demand was for the right to undertake the risks of a clinical trial if he deemed it to be in his best interests. By the mid-1990s the FDA, the pharmaceutical industry, and the patients' rights movement (now joined by people with cancer and other life-threatening conditions) had all converged on the same goal of deregulation.

AIDS activists had tried in vain to draw a sharp distinction between life-saving drugs and more frivolous types of medicine: "We didn't think somebody getting a new hair remedy, Rogaine, should necessarily get accelerated approval of a drug," one prominent advocate recalled, "but certainly people who were facing a life-threatening illness ought to have access to those experimental drugs."[23] For the neoliberals, however, the distinction was without merit. For them, pain-avoidance—in the form of access to experimental drugs on the part of the seriously ill—and pleasure-seeking—in the form of consumer appetite for pharmaceutical rejuvenation—were equally good arguments for dismantling the protections that had been in place since 1962.

In 1997 the Clinton administration enacted far-reaching reforms to the FDA, under the rubric of "modernization." These reforms further loosened the criteria for getting drugs onto the market, shifting much of the burden of testing to pharmaceutical companies themselves, producing conflict-of-interest conditions almost as extreme as those that preceded the thalidomide disaster. They also issued guidelines for direct-to-consumer-advertising of prescription medication, making the United States one of only two countries in the world that allows this form of medical marketing (the other is New Zealand). This move, although controversial, has been repeatedly justified in terms of patient empowerment. According to the nonpartisan National Health Policy Forum, "such information enables the patient to approach his or her physician with information about health status or preferences that might otherwise be lost or overlooked."[24]

THE HEDONIMETER

As the defiant refusenik of the Watergate era metamorphosed into the educated consumer of the new millennium, the discipline of psychology received a matching makeover—courtesy of the economics department. In the 1980s two new disciplines emerged—neuroeconomics and behavioral economics—centered on the pain-pleasure system. Explicitly indebted to Jeremy Bentham, this economic psychology has been so assiduously popularized that it now has some claim to be the consensus view of human behavior. Unchained from the walking wheel of Bentham's Panopticon, Epicurus has been set to work greasing the machinery of global capitalism.

In economic psychology, the road home to Bentham was long and winding. He was a darling of economists in the late nineteenth century but then fell out of favor for many long decades. In 1871 the gold assayer William Stanley Jevons published his influential *Theory of Political Economy*, attempting "to treat Economy as a Calculus of Pleasure and Pain."[25] For Jevons, the value of a commodity could be measured in terms of the pleasure of consumption versus the pain of earning the means to consume.[26] It was Bentham's "thoroughly mathematical" approach that had inspired him to "sum up all the values of all the pleasures on the one side, and those of all the pains on the other."[27] He also paid tribute to Alexander Bain.[28]

The book is best known for Jevons's mathematical analysis of "marginal utility," the way that our appetite for any given commodity is in inverse proportion to the amount of it we have already consumed. For our purposes, what is interesting is Jevons's assumption that people were engaged in a constant quest to maximize pleasure and minimize pain. From this foundation he raised a whole science, in which the economic activities of "buying and selling, borrowing and lending, laboring and resting, producing and consuming" manifest the pains and pleasures of the human organism.[29] "Almost unconsciously we make calculations of this kind more or less accurately in all the ordinary affairs of life," he declared.[30] For Jevons, the measurable facts of economic activity were symptoms of human misery and delight, an assumption about the psychological underpinnings of getting and spending that required no further explication.

Jevons was pessimistic about the possibility of any sort of direct measurement of pain and pleasure, but his immediate successor, the economist Francis Edgeworth, refused to give in to these technical shortcomings. In a short book called *Mathematical Psychics*, Edgeworth proposed that although "hedonism may still be in the state of heat or electricity before they became exact sciences," the measurement of pain and pleasure would one day be possible. To this end, he conjured up an imaginary "psychophysical machine" that he called the "hedonimeter":

> From moment to moment the hedonimeter varies; the index now flickering with the flutter of the passions, now steadied by intellectual activity, low sunk whole hours in the neighbourhood of zero, or momentarily springing up towards infinity. . . . Any just perceivable pleasure-increment experienced by any sentient at any time has the same value. . . . We have only to add another dimension expressing the number of sentients, and integrate through all time and over all sentience, to constitute the end of pure utilitarianism.[31]

"These are no dreams of German metaphysics," Edgeworth reassured the reader, "but the leading thoughts of leading Englishmen, and cornerstone conceptions, upon which rest whole systems of Adam Smith, of Jeremy Bentham, of John Mill, and of Henry Sidgwick."[32] His patriotic hedonimeter would solve all sorts of political, moral, and economic questions, from political equality to Irish land reform, by measuring "what may be dimly discerned as a sort of hedonico-magnetic field."[33]

In the absence of real hedonic machinery, Edgeworth's musings were dismissed. If economics was going to be a hard science, practitioners argued, it could not rest on the foundation of psychology. Introspection was the only way to gain access to subjective states, and introspection was unscientific. Anyway, the sources of delight or misery varied too much between individuals. The roads to happiness taken by the ascetic monk, the brave soldier, and the ordinary bourgeois householder could never be made equivalent.[34] By the first decades of the twentieth century, economists had stopped referring to the pain-pleasure underpinnings of economic behavior.[35] Even the word "utility" was judged confusing: "The term is a heritage of Bentham and his utilitarian philosophy. It is misleading to every beginner in economics and to the great untutored and naïve public, who find it hard to call an overcoat no more truly useful than a necklace."[36]

After World War II, things began to look even worse for any link back to utilitarian psychology. For Cold Warrior economists such as Friedrich Hayek, political freedom and free markets were inextricable from the glorious metaphysics of free will. Famous works such as Hayek's *The Road to Serfdom*, published in 1944 but reaching millions more readers when a condensed version appeared in the *Readers' Digest* right at the end of World War II, made the case against utilitarianism in the same terms as Arthur Koestler's *Darkness at Noon*. Here is Hayek on the utilitarian ends-means relation in totalitarian societies:

> The principle that the end justifies the means, which in individualist ethics is regarded as the denial of all morals, in collectivist ethics becomes necessarily the supreme rule. There is literally nothing which the consistent collectivist must not be prepared to do if it serves "the good of the whole."[37]

Hayek's critique of socialist utilitarianism was as much a theory of sacred autonomy as anything coming out of the antipsychiatry movement.[38] Bentham and Bain's place in economics seemed to be completely obsolete.

In 1979, however, a new discipline appeared on the horizon, restoring pleasure and pain to the beating heart of economic theory. Coincidentally or not, the breakthrough occurred just as Deng Xiaoping, Margaret Thatcher, and Ronald Reagan came to power in rapid succession. "Behavioral economics," as this field is known, started out as a challenge to the prevailing wisdom of rational-choice theory in mainstream economics. Rational-choice economists made predictions based on the assumption that people want always to maximize their economic gains and minimize their losses, a vision of humans as calculative machines delinked from any actual theory of human nature. In 1979 the Israeli psychologists Daniel Kahneman and Amos Tversky published an article refuting this assumption. The idea that "all reasonable people" would wish to obey utility maximization and that "most people actually do" seemed to them to be unsupported by the evidence.[39] For example, they ran a series of experiments in which they found a consistent difference between people's behavior when confronted with the possibility of gaining an amount of money or losing the exact same sum. While the idea of loss aversion may not seem particularly world-changing to a non-economist ("When they are giving, take; when they are taking, scream," as my father-in-law used to say), this insight proved transformative for the discipline.

Holding out the promise of more accurate predictions of consumer behavior, Kahneman and Tversky started to urge economists to pay attention to real-world evidence for how human beings actually make decisions. They called their new framework "prospect theory" (a deliberately meaningless moniker), and it made an enormous impact: the 1979 paper has been cited tens of thousands of times, and was decisive in Kahneman's winning the Nobel Prize for economics in 2002 (Tversky died in 1996).[40] Above all, because prospect theory was empirical rather than formalistic, it lent itself to naturalistic explanations from evolutionary theory. This new breed of economists surmised that human deviations from perfect rationality might be explained by how the evolution of our species gave rise to the quirks of our nervous systems.

With its emphasis on naturalistic explanations for utility-seeking behavior, behavioral economics has beaten a path back to all the questions about utilitarianism that animated British intellectuals at the end of the nineteenth century. In 1997 Kahneman coauthored a paper entitled "Back to Bentham?" appealing to Edgeworth's hedonimeter as a precedent for his own attempts to track pleasure and pain. His musings on utilitarian psychology stemmed from an intriguing experiment. He had asked colonoscopy patients to report how uncomfortable the experience was as it was happening, and then to rate its unpleasantness afterward. He found that after the procedure was over people consistently rated their discomfort in terms of an average between the most painful stretch and the very last part of the experience. Kahneman formalized this into what he called the Peak-End Rule: our memories of pain and pleasure are determined by the period of the greatest intensity and the last period. It turned out that people would rather undergo the repetition of a longer colonoscopy with a more pleasant ending, than a shorter one that ended nastily, even though they had endured more aggregate discomfort in the former based on a moment-by-moment report of the experience.[41] Kahneman suggested that we should use such data about "experienced utility"—reports of how delighted or miserable an individual is at any given moment—as a tool for correcting this irrational preference for more pain overall. His updated hedonimeter was not a piece of instrumentation; it was a conceptual commitment to real-time recording of subjective states.[42]

The cultural ramifications of this approach are spreading fast. Real-time hedonic self-recording has become part of everyday life in the information age, from the ubiquitous thumbs-up icon, to the stream of consumer reviews and recommendations, to the "quantified self" movement, devoted to the continual tracking and analysis of personal data. "Life-logging," as this last activity is known, has its most salient and popular applications in healthcare—people are monitoring chronic conditions such as asthma and epilepsy in order to track triggers, and logging exercise and nutrition data to help with weight loss. The emergence of a new breed of empowered patients keeping track of moment-by-moment sensations on their smart phones marks the frontier of consumer empowerment in medicine, reconciling experienced utility and patient autonomy within the framework of economic individualism.

REASONING DUCKS AND RATIONAL ADDICTS

While Kahneman and Tversky labored to undermine rational-choice theory, another school of economic behaviorists heralded the triumphant return of pleasure-pain psychology in a more overtly Skinnerian key. Over the course of the 1990s, in the wake of the collapse of the Soviet

Union, the relation between economic activity, pleasure, and biology became the subject of an interdisciplinary field called neuroeconomics. In a 1993 collection of essays on neuroeconomic themes, the psychologist Richard Herrnstein (coauthor of the controversial 1994 book about inherited intelligence *The Bell Curve*) credited Bentham with giving "such impetus to what he called the Principle of Utility that, two centuries later, it continues to be the flywheel of behavioral and social science, as well as of political, moral and legal philosophy."[43] The article ended with the thought that "we descendants of Bentham can agree about the primacy of pains and pleasures."[44] The author presented this Benthamite tradition as unbroken, but he was actually in the vanguard of a revival, coinciding with the end of the Cold War. The philosophy that had been dismissed in economics, denounced in psychology, and demoted in medicine was returning to intellectual respectability.

One representative neuroeconomic article, published in 2004, described experiments with humans, ducks, and chimpanzees. The human subjects were given a classical strategic game to play, in which an "employee" had to decide whether to "work or shirk" and an "employer" had to make up her mind to "inspect or not inspect" (a panoptic setup of which Bentham would surely approve). Each of these decisions came at a preset cost, and the researchers tracked how quickly the players established a utility-maximizing strategy under different conditions. This human behavior was compared with the behavior of mallard ducks when two people threw different amounts of bread at them, and a series of experiments in which thirsty monkeys were given variable amounts of delicious fruit juice in return for performing particular actions.

The ducks were fed bread balls of different sizes at different intervals by two different researchers. It turned out that the ducks quickly arranged themselves in front of the researchers in such as way as to maximize their chances of grabbing a bite to eat. When the bread-throwing rate was altered, the ducks rearranged themselves into a "perfect equilibrium" within *seconds*. Impressed, the authors concluded that "the ducks as a group behaved in a perfectly rational manner, in a manner that many economists would argue was evidence of a rational conscious process if this same behavior had been produced by humans operating under simple market conditions like these." The last sections of the paper moved into neurochemical territory to explain this result. They posited an area of the brain that computed "expected utilities," based on inputs from the dopamine circuits, the feel-good neurotransmitter that is thought to mediate reward. The authors concluded that "economics and biology are two disciplines addressing a single subject matter. Ultimately economics is a biological science. It is the study of how humans choose. That choice is inescapably a biological process."[45]

Compare this to a study Skinner published in 1937. He had heard about some experiments in which chimpanzees "earned" poker chips and "spent" them on food and other treats. These behaviors were presumed to be higher intellectual functions, beyond the reach of nonprimates, so Skinner promptly set himself the goal of teaching a rat to "spend money." Breaking the task down into small units, each of which he rewarded, he managed to get one of his rats to pull a chain to release a glass marble, pick the marble up, carry it across his cage, and drop it into a tube, in exchange for food. He noted that the subject, when not hungry "would occasionally pull the chain to release marbles which he did not immediately spend, though I doubt whether he was 'hoarding money for its own sake.'"[46] Skinner reduced humans to conditioned rats; the neuroeconomists inflate mallard ducks to economic humans. In both cases, "rationality"—an illusory quality for Skinner and a ubiquitous one for neuroeconomists—is assumed to be no more than the operation of a neural feedback circuit connecting pleasure and behavior.

Around the same time, the related specter of the "rational addict" began to stalk the halls of the business school. In 1988 two influential conservative economists proposed "a theory of rational addiction in which rationality means a consistent plan to maximize utility over time."[47] Their definitions of both addiction and rationality were completely Skinnerian: "A person is potentially addicted to c," they announced, "if an increase in his current consumption of c increases his future consumption of c. . . . Reinforcement means that a greater current consumption of a good raises its future consumption."[48] The paper showed how this relation between current and future consumption—"drug tolerance" to you and me—could be mathematically modeled. Although it was necessary to assume steep indifference to future happiness in the calculations of the addict, the model still portrayed taking heroin as a rational activity. The stated goal was to predict the outcome of policy initiatives such as tobacco taxes. By interpreting the heroin addict as a reasoning machine, these theorists pulled off the same trick as the neuroeconomists—a complete assimilation of rationality into pleasure-seeking and pleasure-seeking into rationality.

During the Cold War, a notion of the relation between moral freedom, political freedom, and free markets kept economists away from the rats and pigeons of the behaviorist laboratory. Free choice lay at the heart of the ideological argument for market economies, and behaviorism's insistence on psychological determinism was incompatible with the struggle against communism. With the collapse of the Soviet Union, the prohibition melted away like snow in spring.[49] According to neoliberal psychology, all sentient creatures are stimulus-response mechanisms, embedded in markets for things they desire—money, bread balls, sips of juice, food rewards, or heroin. With the Cold War over, Friedrich Hayek's beloved *Homo economicus*—personification of political liberty, autonomy, and free will—became just another rat pressing the pleasure lever.

WHY THE TORTOISE WINS THE RACE

Reasonable ducks and rational addicts are unitary creatures in whom the drive for pleasure is everything, but the currently most influential form of economic psychology splits desire and deliberation into two neural systems. "Dual process theory" (DPT) argues that we have two different ways of making decisions: a fast, intuitive one and a slow, deliberative one. The intuitive mode is disparaged as primitive and error-prone; the deliberative mode is celebrated as rational and farsighted. The former makes you gobble down the chocolate cake; the latter instructs you to choose the salad. If you are one of the lucky ones, the slow, deliberative choice will win out over the primitive impulse. Calculative reason is therefore the key to achievement in most areas of life, from losing weight to climbing the career ladder.

The connection between deliberative thinking and worldly success is popularly represented by the "Stanford marshmallow test," a psychological experiment that has become one of the fables of social mobility in the neoliberal age. The marshmallow test has its rather unlikely origins in an experiment conducted in the late 1950s as a response to postwar anxieties about racial prejudice. In 1957 Walter Mischel, a new-fledged professor of psychology at the University of Colorado and the son of

Jewish refugees from Nazi Europe, went on a trip with his wife to the island of Trinidad. At the time the island was still under British rule, but independence was on the horizon, and the brilliant black historian Eric Williams, author of the classic *Capitalism and Slavery*, was chief minister. Unsurprisingly, racial tensions were running high. When Mischel arrived, "numerous informants" told him that "the Negroes are impulsive, indulge themselves, settle for next to nothing if they can get it right away, do not work or wait for bigger things in the future," while the "Indian is said to deprive himself and to be willing and able to postpone immediate gain and pleasure for the sake of obtaining greater rewards and returns in the future."[50] Mischel set himself the task of dismantling this negative stereotype.

To do this, Mischel appealed to a body of work linking a person's ability to delay gratification to *trust*. In his analysis, context was all-important. It made sense for someone to hold out for a bigger treat only if they believed it would actually appear when promised. In a more capricious and unreliable world, a rational being would take what was on offer in the moment. Steeped in the soft-Freudianism of 1950s American psychology, Mischel connected stable reward systems to family life and hypothesized that the supposed racial difference between black and Asian children was in fact due to the presence or absence of a father figure at home. When promised a future reward by a male researcher, he surmised, those children with a father at home would be more likely to believe that the delayed treat would really manifest, and therefore be more willing to forgo the smaller, cheaper candy right under their noses, regardless of race. Without faith in the system of reward delivery, there was no reason for a child to give up the pleasure right in front of him.

To test his hunch, Mischel took a group of children ages seven to nine out of a Trinidad school, divided them up into the two groups based on race, and had them fill out a questionnaire about their home lives. He then told them that he wanted to give them some candy to thank them for their efforts. Unfortunately, he confessed, he only had small, cheap candy on hand, but if they were willing to wait a week, they could have a bigger, more expensive sweet. Just as he expected, he found the willingness to delay gratification to be positively correlated with the presence at home of a father figure. It also seemed to be strongly age-dependent. He concluded his article with the thought that there might be more interesting work to be done on the phenomenon.

A decade later, Mischel followed up the study, this time with children from three to five years old at a Stanford University nursery school. In this version there was no racial, cultural, or economic element to the analysis. The subjects were the children of Stanford faculty and were assumed to be socially homogeneous. Mischel's goal was to see what experimental conditions might favor longer gratification delays. For example, did being able to *see* the delayed reward help the child to resist the treat in front of her? If she could see two marshmallows glowing behind a window did that boost her motivation to resist the single one within reach? The answer to that question was a resounding "no." The opposite was true. Famously, the children who did best at the task devised "elaborate self-distraction techniques" such as covering their eyes, talking to themselves, singing, or playing games with their hands and feet.[51]

Mischel went on testing children in this way until eventually more than six hundred Stanford preschoolers had undergone some version of the experiment. A decade or so later, he mailed questionnaires out to all their parents (at least all the ones he could trace), asking a series of questions about his research subjects, who were now adolescents. It turned out that those who were better able to delay gratification as very young children were more likely to be able to cope with the stresses of high school and even had higher SAT scores. In vain did Mischel and his coauthors "emphasize the need for caution in the interpretation" of these findings.[52] The study and its follow-up questionnaire has become one of the most overinterpreted experiments in the annals of psychological science, as well as spawning a popular subgenre of YouTube videos featuring preschoolers squirming in front of marshmallows. How you do as a four-year-old on the "marshmallow test" is now taken to be potentially predictive of your whole life course.

One of the scholars who took these results and ran with them was the behavioral economist Richard Thaler. In 1981 Thaler and a collaborator published a paper that they described as "the first formal treatment of two-self economic man," divided into "a farsighted planner and a myopic doer."[53] The struggle to refrain from eating a marshmallow was represented as a war between opposing impulses in the human breast. Thaler would be the first to admit that the planner and the doer were no more than metaphors, but they were the right metaphors at the right time. Before long a slew of variations on the theme had appeared. The planner in all these schemes corresponded to the rational agent of traditional economics, able to maximize utility with a good grasp of logic, mathematics, and the relevant facts; the doer corresponded to the irrational protagonist of the new behavioral economics, prone to make investment mistakes, to overeat, and not to save enough for retirement.

In 2008 Thaler and his colleague Cass Sunstein published *Nudge*, a best seller that popularized dual process theory and made think-tank stars of its authors. In the book, they urged that the insights of behavioral economics be taken out of the seminar room and into the realm of public policy. The idea is that because people are error-prone, they need to be "nudged" in the direction of making good decisions. This is achieved by setting up the "choice architecture" in such a way as to encourage sensible behavior. Opt-out rather than opt-in schemes for retirement savings are the most obvious example. Thaler has dubbed this strategy "libertarian paternalism."

Libertarian paternalism has been a huge hit. The Tories in Britain and the Democrats in the United States, to mention just two, have integrated it into policy design. DPT is the ideal psychological theory for the Anglo-American "third way" between unfettered neoliberalism and old-fashioned social democracy. David Cameron had a "nudge unit," Barack Obama a behavioral insights team. A spin-off from the British team works with the Australian government. Trials of nudges are being conducted in every corner of the globe.[54] At some point Daniel Kahneman made the connection between dual process theory and his own work on irrational behavior, and in 2011 he published his smash best seller *Thinking, Fast and Slow*, explaining the psychology of the two selves. Part 5 of the book opens with a tribute to Bentham.[55]

TROLLEYOLOGY

In Kahneman's paper "Back to Bentham?" the question mark in the title hovered over utilitarian ethics. Kahneman is a dedicated investigator of the motivating force of pain and pleasure, but he stops short of the view that utilitarianism provides an exhaustive standard of right and wrong.[56]

Recently, however, dual process theory has taken a turn toward full-bore utilitarian ethics of a really quite extreme variety, courtesy of Harvard neuroscientist Joshua Greene. In 2013 Greene published a devoutly Benthamite treatise entitled *Moral Tribes*, which has touched off a debate eerily similar to the one that raged about the nature of the moral sense in nineteenth-century Britain.

Greene designates DPT's two modes of decision-making as "automatic" and "manual," by analogy with a camera. The automatic mode allows you just to point and shoot; the manual mode forces you to choose all the settings yourself. The automatic system produces moral intuitions such as "don't lie." The manual system produces moral calculations such as "tell a lie in this situation because it will be better for everyone involved." Although Greene seems to be unaware of his distinguished Oxford predecessor, the argument of *Moral Tribes* is a faithful recapitulation of R. M. Hare's *Moral Thinking: Its Levels, Methods and Point* (1981), the book that turned me into a card-carrying utilitarian at age eighteen. Hare's two levels of moral thinking—the "archangel" who knows everything and calculates consequences, and the "prole," who knows nothing and relies on moral intuitions—exactly correspond to Greene's manual and automatic modes. It was slightly uncanny to be finishing the first draft of this book, only to be confronted by a complete restatement of the philosophy that set me off on my delinquent utilitarian path in the first place.

Of course, there are differences between the earlier and later works. In contrast to Hare's elegant analysis of moral language, Greene's research is distinguished by its absolute literal-mindedness. For him, dual process morality is located in the material brain. The automatic mode—primitive and irrational—is a deep brain structure; the manual mode—highly evolved and calculative—is located in the frontal lobes. Hoping to "use twenty-first century science to vindicate nineteenth-century moral philosophy against its twentieth-century critics,"[57] Greene put people in brain scanners, and asked them to ponder a classic utilitarian thought experiment about an out-of-control trolley, a research practice he self-mockingly calls "trolleyology."

The "trolley problem" offers two scenarios. In the first you have to decide whether to pull the switch that will divert a runaway trolley from its course, thereby saving the lives of all five people aboard but causing the death of one person who will be caught in the trolley's new path. The other scenario asks you to decide whether to push someone off a footbridge to block a runaway trolley's path, thus saving five lives at the cost of one. (Greene, a witty and approachable writer, has done the world of philosophical ethics a service by tactfully updating the unfortunate victim of this scenario from being "fat" to "wearing a big backpack," which explains why only the person standing next to you can do the job of blocking the train, and you can't jump off the bridge yourself.) People tend to say yes to the first scenario and no to the second, although the utility calculus is the same in both cases. Greene asserts that in the second scenario the emotional stress of having to imagine actually pushing someone off a footbridge gets in the way of the correct utilitarian answer.

The moral squeamishness we feel about committing murder in the second scenario, Greene claims, has been embedded in our nervous systems by evolution. This does not, in his view, make it right. He proposes that our intuitive rules of conduct—don't kill, don't lie, don't steal—evolved for the purposes of what he calls *in-group* cooperation. Following these rules is the way to get along with the other members of your tribe. But, he asserts, these cooperative tendencies were able to evolve only because they produced a competitive advantage *between* different groups: "The only reason natural selection would favor genes that promote cooperation is that cooperative individuals are better able to outcompete others."[58] Cooperative instincts are therefore nothing to be proud of. They are fatally limited by their evolutionary origins. Our repugnance at the idea of committing a violent act, even for the sake of the greater good, is simply a primitive, tribal mode of moral reasoning that we need to evolve beyond.

But, Greene reassures us, we have the capacity to transcend our tribalism, if only we would press our prefrontal cortices into service and do a little "manual" moral thinking: "If I'm right, Bentham and Mill did something fundamentally different from all of their predecessors, both philosophically and psychologically. They transcended the limitations of commonsense morality by turning the problem of morality (almost) entirely over to manual mode":

> The manual mode's job is, once again, to realize goal states, to produce desired *consequences.* . . . In other words, the human manual mode is

designed to produce *optimal consequences*, where "optimal" is defined by the decision-makers' ultimate goals, taking into account the expected effects of one's actions.[59]

Again, Greene's manual and automatic modes correspond exactly to the "archangel" and the "prole" of Hare's utilitarian creed: two levels, one calculative, utilitarian, and heavenly, the other intuitive, primitive, and implicitly lower class. But Greene's denigration of "tribal" intuitions is a far more extreme argument for the supremacy of the archangel than anything advanced by Hare.

THE VOICES OF MILL AND KANT

One of Greene's predictable targets is the notion of human rights. Like Bentham, Malthus, and Skinner before him, he finds the concept incoherent and self-defeating: "If we are truly interested in persuading our opponents with reason, then we should eschew the language of rights."[60] Greene avers that invoking human rights sometimes serves when reason fails, but that they are completely expendable in the name of the greater good.

Greene's views on rights and utility are nicely exemplified by a study he conducted using medical professionals as the subjects. He hypothesized that medical doctors would tend to be individualistic in their moral judgments while public health professionals would tend to be more utilitarian. Sure enough, the public health professionals were more likely to say "yes" to shoving the poor guy with the backpack off the bridge and also "to approve of rationing drugs, quarantining infectious patients, et cetera." The doctors, by contrast, clung to a notion of the autonomous individual who should never be used as a means to an end. For Greene, this was a perfect example of dual process morality at work in "the real world of healthcare decision-making." In his terms, the physicians in his experiment were afflicted with unreasonable tribal morality, while public health professionals were shining exemplars of rational, manual-mode thinking.

The respondents in the medical study were invited to add comments to their answers, from which Greene picked out two particularly salient examples:

One public health professional wrote, "In these extreme situations . . . I felt that a utilitarian . . . philosophy was most appropriate. Ultimately, that is the most moral thing to do. . . . It seems the least murky and the most fair." In contrast, a medical doctor wrote, "To make a life-or-death decision on behalf of someone who is capable of making that decision for themselves (and who has not forfeited that right, for example by knowingly committing a capital crime) is a gross violation of moral and ethical principles."

Greene glosses this passage as "The voices of Mill and Kant, speaking from beyond the grave."[61] He goes on to muse on the question of cause and effect, wondering if utilitarian types are particularly prone to go into public health, or if their public health training turns them into utilitarians. He never asks why doctors might speak so passionately and clearly with the "voice of Kant." And, indeed, why should he? Medical ethics has not made its case against utilitarianism with the force and clarity that history demands.

Kant is a kind of joke for Greene, who has some fun at the expense of his arguments against masturbation. But I hope that this book has shown that the Kantianism of the medical profession is not so much an example of crude tribal morality as an indispensable political achievement. To use R. M. Hare's language, all medical professionals used to be archangels of utility who knew what was best for the rest of us. In the age of penicillin and the polio vaccine the utilitarian calculus seemed infinitely weighted in their favor, especially when it came to enrolling disenfranchised people into medical research. After the Eichmann trial, however, the lessons of history were too raw and recent to ignore the fact that the archangels were violating the Nuremberg Code. The resignation of the three Jewish physicians who refused to inject cancer cells into their patients lit a fuse that eventually exploded with the revelation of the Tuskegee syphilis study. As a result, since the early 1970s, doctors and researchers have been imbued with the pragmatic Kantianism of informed consent protocols, and their training has increasingly stressed the importance of patient autonomy.

The "voices of Kant and Mill" may indeed correspond to two modes of reasoning that have distinct neural substrates, just as Greene would have us believe, but what those voices say is the product of a history whose lessons should not be squandered. From Jeremy Bentham to Joshua Greene, utilitarians have argued that rights are justified not on their own merits but only insofar as they might be productive of happiness. But as I hope this narrative has shown, the right to informed consent is a principle *explicitly formulated to trump utility considerations*. The four principles of biomedical ethics comprise not just a list but a *ranking*: autonomy, nonmaleficence, beneficence, and justice. As the philosopher and legal theorist Hugo Adam Bedau eloquently put it:

> Far from being the product of a calculus of pleasure and pain, as our legal rights are on Bentham's theory, the very nature of human rights is to be the enemy of such calculations. Indeed, it is precisely because of the assault those calculations often require against individuals that we rely on human rights to obviate them.[62]

Utilitarianism may have reason on its side, but the historical record suggests that we are not built for it. Greene and all proponents of dual process theory insist that the archangel and the prole coexist in every human breast. But what purports to be a description of two levels of decision-making always threatens to tip over into two classes of decision-maker: a small cadre of scientific experts licensed to make decisions on behalf of a suffering humanity, and the great mass of their less educated, less reasonable, less qualified beneficiaries.

ARCHANGELS AND DEMIGODS

Fortunately, dual process theory is not completely lost to Bentham. There is a more promising story floating around about its British paternity. Many behavioral economists credit Adam Smith with identifying the two systems of decision-making. "Smith said it first," confesses Richard Thaler, coauthor of *Nudge*. According to Thaler and some other behavioral economists, their "rational planner," the eschewer of single marshmallows who sticks to the salad and saves her money for a rainy day, is the same as Smith's "impartial spectator" from *The Theory of Moral Sentiments*.[63]

The rational planner of modern DPT does, indeed, have a place in *The Theory of Moral Sentiments*. Smith describes it in the same terms used by behavioral economists today:

> The qualities most useful to ourselves are, first of all, superior reason and understanding, by which we are capable of discerning the remote consequences of all our actions, and of foreseeing the advantage or detriment which is likely to result from them: and secondly self-command, by which we are enabled to abstain from present pleasure or to endure present pain, in order to obtain a greater pleasure or to avoid a greater pain in some future time.[64]

While this passage constitutes as good a description of the behavioral economists' "system two" as you are likely to find, it is emphatically not about Smith's "impartial spectator." Smith calls the ability to delay gratification in pursuit of long-term gain the "virtue of prudence," and it plays only a very minor role in his scheme.

Smith's "impartial spectator" is a far richer, deeper, and more subtle concept. He sometimes calls it "the demigod within," by which he means our innate ability mentally to switch places with other people and see the world from their perspective. It arises from our inborn capacity for sympathy. It is explained by the logic of our existence as social and rational animals. It comprises the emotional basis of ethical action. Like prudence, it springs from reason, but unlike prudence it is primarily directed toward the views and interests of others. The idea that it is the same as the rational planner of dual process theory would be laughable if only behavioral economics wasn't so influential. If you're going to endorse Adam Smith, endorse Adam Smith, not Richard Thaler in a powdered wig.[65]

The Theory of Moral Sentiments partakes of all the earthy wisdom of utilitarianism's stress on pleasure, comfort, and security, without ever confusing us with mere sacks of sentience. For Smith we are sentient *and* rational. We feel *and* we think. We are not rats in Skinner boxes. (Hell, not even *rats* are rats in Skinner boxes.) Nor are we shining exemplars of Kantian propriety, answering only to the bloodless imperatives of duty. Above all, we are social beings, with an inborn appreciation for the grammar of reciprocity. "Do unto others as you would have them do to you" is, indeed, the most basic rule of moral life, and by cultivating Smith's demigod within we can improve upon our practice of it.

Smith's analysis of moral reciprocity explains why informed consent compels such wide, deep, and intuitive agreement. When people learn that there was a time when researchers did not adhere to its protocols, they invariably express shock and surprise. Informed consent is not just a civil right that happens to be applicable in medical settings; it is an exchange with a profound ethical structure. It confers the privilege of self-determination on the patient; it demands from the doctor the discipline of respect. Explaining what a patient needs to know about a procedure, and then honoring her assent or refusal, is a conversation conforming to the fundamentals of humanist morality, including the intellectual effort to see the situation from another's perspective, to have true regard for her freedom, and to level the asymmetry of power and expertise in order to meet her on more equal ground. The rise of informed consent represents a triumph for Smith's exact brand of Enlightenment morality. His impartial spectator is a habit of thinking from the perspective of others; informed consent is its practical arm. Don't assume you can work out on your own how others would like to be treated. Find out from them!

As Smith acknowledged, there is no dodging the fact that we all have to engage in deliberative, calculative, consequentialist reasoning sometimes. Public health administrators have to make these kinds of calculations *all* the time. This is why their decisions are called "tough." Epidemic conditions also demand such utility calculations, as in the preferential administration of scarce antiviral medications to health workers at the height of the Ebola outbreak. But to elevate this emergency type of reasoning over a baseline respect for human rights is to throw away the hardest-won moral lessons of the last seventy years. If we have learned anything, it is surely that the archangel must climb down to earth and meet the rest of us face to face. As soon as he thinks he reigns supreme, his Miltonic fall is inevitable. It is the humble prole, the believer in the golden rule of moral interchangeability, a denizen not of heaven or hell but of our own imperfect world of communication and compromise, who alone can be trusted with the sovereign master's throne.

The Restoration of the Monarchy

Sitting here in my adopted home of California, banging out a book about British philosophers, I sometimes feel like an obscure Greek scribbler in a distant outpost of the Roman Empire, a decadent representative of the old and exhausted culture that once gave birth to the society around me. It is no coincidence that I converted to the principle of utility at Oxford University. That beautiful city has always nurtured utilitarians. R. M. Hare taught at Balliol College. As a student, I unknowingly ate grilled-cheese sandwiches in the same Queen's Lane coffeehouse where Jeremy Bentham alighted on the greatest happiness principle. Hobbes matriculated at my own college at the tender age of fifteen. A neo-utilitarian movement flourishes there now under the rubric of "effective altruism." It is a haunted place. The centuries accordion together within those ancient walls. The charisma of Oxford's golden stone, its High Tory leanings, the archangelic self-confidence of High Table—these are all braided into the very sinews of the utilitarian self.

The day in 1983 when I caught the Oxford disease and became a utilitarian, I walked back to my room at Hertford College (Magdalen Hall in Hobbes's time) and flung myself on the bed to ponder the implications of my conversion. Hare's book had talked a lot about moral intuitions. He was not as scathing about these as his Harvard successor Joshua Greene would be; they played an important role in his scheme. He also seemed to take for granted that everyone had them. I was not so sure. I remembered moral intuitions from my childhood, but it was a while since I had had one. Did I, I asked myself, *have* any moral intuitions? Or, more precisely, did I have any moral intuitions *left*, after spending the last few years in a fog of teenage nihilism?

Searching my soul, I found that there was a moral intuition lurking there, rather a strong one as it happened. "Save the trees," an inner voice pleaded. "Saving trees is good." As moral intuitions go, this one, it turned out, was easy to translate into utilitarian terms. What could be more conducive to the greater good, after all, than the preservation of the natural resources on which all life and pleasure ultimately depend? Although I was to become pretty eclectic in my activism ("rent a mob," my mother once called it in mock despair), in some ways the environment remained at the heart of my utilitarian calculations. Not only should the few be sacrificed to the many and the rich to the poor, my inner archangel urged, but at an even more fundamental level all the reckless First World wasters of our fragile planet should be thwarted, by any means necessary, for the sake of future generations. Hare had turned me into a militant Malthusian.

Of all the legacies of the utilitarian reform period in British history, perhaps none is as toxic as that of Thomas Malthus. Old Pop was a humane individual, but his conclusions were deeply antihumanist. His philosophy implied that human increase was the source of famine, illness, misery, and war. Ever since his first appearance on the world stage, his views have been an incentive to snobbery and racism. In accordance with his extreme consequentialism, his followers dismantled the existing system of welfare relief. John Stuart Mill loathed poor people to the end of his days because of his unbending loyalty to Malthus. In the 1880s Malthus's population principle metamorphosed into racist eugenics. The fascist concept of *Lebensraum*—the idea that the German people needed a vast agricultural hinterland in order to expand and prosper—was directly in the Malthusian tradition. Even in our own supposedly postcolonial age, Malthusian doom-mongers sometimes exhibit disproportionate anxiety about the teeming masses of the global south.

Rather like John Stuart Mill, I find I have a hard time letting go of Old Pop's remorseless logic, despite knowing how unpleasant his disciples proved to be. Even as I have abandoned other facets of greater-good thinking, environmental anxiety remains a piercing rebuke to my evolving bourgeois ethics. I suffer from a form of quiet madness that a friend aptly calls "eco-pathology": an obsessive and futile quest to minimize my personal environmental footprint.[1] Wobbling along the roads of San Diego on my bike, in constant fear of being flattened by the gas-guzzling behemoths sweeping past me, my inner Malthusian misanthrope mutters furiously on, albeit *sotto voce*, as befits the respectable middle-aged woman I seem to have become.

But luckily for all of us, Malthus was wrong. He was wrong about the inevitability of starvation on his island nation. He was wrong about the limits on agricultural yields. He was wrong about the ineradicable brutishness of human life. Instead, the palm of prophetic accuracy has to go to the optimists who opposed him, to those anti-Malthusians who believed in human ingenuity, universal rights, the abolition of slavery, and the sweet promise of science. The same year as Malthus published his *Essay on the Principle of Population*, Edward Jenner published his *Vaccination against Smallpox*, a tract that made such a material difference to the Malthusian predicament that Charles Darwin himself invoked it in a passage about the thwarting of natural selection. "There is reason to believe," Darwin shuddered in one of his more eugenic moments, "that vaccination has preserved thousands, who from a weak constitution would formerly have succumbed to small-pox."[2]

Consider Jenner's friend Thomas Beddoes, the Epicurean polymath who founded the Pneumatic Institute in Bristol where the laughing gas trials were staged. In 1799 Beddoes published a collection of observations and experiments penned by his West Country circle, including Jenner. The volume included a report by Humphry Davy on the oxygenation of the blood. Jenner made a plea for what would now be called evidence-based medicine. Treatments for gonorrhea took up many pages without a trace of moralizing. In his introduction Beddoes declared that all systems of morality ultimately reduce to the "well-being of individuals." "How," he demanded, was "any progress in genuine morality" possible without "accurate ideas of the causes that affect the personal condition of mankind"?[3] In other words, the greatest contribution to moral progress would issue, ultimately, from scientific medicine.

Who among us could sincerely repudiate Beddoes's creed? Anyone who has ever gulped down an antibiotic is by implication a believer—or a hypocrite. Is there a reader of this book who can, without contradiction, cast the first stone against the miracles of medical science? Beddoes was a philosopher of health and happiness, but he was no utilitarian. He was a friend of the French Revolution and a disseminator of Kantian ethics in Britain. Malthus be damned, he exhorted all practitioners of medicine in "charitable institutions for the indigent sick," such as his own Pneumatic Institute, to pool their knowledge in order to improve their skill at keeping the most unfortunate members of the polity alive and well. Imagining the burgeoning of human happiness through the promise of science, Beddoes and his friends kept aloft Joseph Priestley's Enlightenment optimism long after the master was exiled to Philadelphia.

Priestley's brand of Enlightenment optimism seems to be on the upswing these days. I recently watched the 2015 film *The Martian*, in which Matt Damon plays the title role, an astronaut left behind on Mars. He unexpectedly survives the dust storm that was assumed to have killed him and then has to figure out how to grow enough food to last him the four earth years until he can be rescued. I could not help but read the story as an optimistic parable of survival on our own warming planet. Damon—a white guy, of course, but one who knows how to take orders from a female captain—is a botanist by training, and he grows a potato crop in Martian soil fertilized by the feces of the departed crew. The potato seedlings sprout with hairy vigor in a big white tent, scenes that draw deeply on the aesthetic of urban farming, the extraterrestrial version of beekeeping on city rooftops. Damon's Martian is a Brooklyn hipster in a space suit, with indomitable green fingers that will save the world. He's also a direct scientific descendant of Priestley, Davy, and the laughing gas revelers, and cutely blows himself up trying to make water from hydrogen and oxygen (not, according to one aerospace engineer, the most efficient way to hydrate on Mars).[4]

The film is utterly ridiculous, and yet it just manages to lick at the shores of scientific plausibility. Its message? That we can and we will, as Damon puts it in one memorable speech, "science the shit" out of our environmental predicament. After nine years at the University of California, home of cutting-edge climate science and visionary bioengineering, who am I to disagree? It's not that I have ceased to worry, but just that I have begun to harbor a tiny ember of hope that worrying long and hard enough about global warming as a scientific and technical problem will actually get us somewhere. Scientists at the J. Craig Venter Institute down the road, for example, recently embarked on an ingenious project to engineer a blue-green alga capable of producing hydrogen as an energy source.[5] In principle, the organism would make the hydrogen from water, and water would be the only by-product of its use as a fuel. They had to abandon the experiments when falling oil prices caused the Department of Energy to pull the funding.[6] But if the day ever comes for this solution (or something like it), it will be the fulfillment of everything Joseph Priestley imagined would be possible when he worked out the basic chemistry of photosynthesis.

From what I know of Priestley and his circle, there's one solution to exponential human increase that they would have endorsed with the greatest enthusiasm, and that is the empowerment of women through reproductive choice, employment opportunities, and education. Of all the strategies for stabilizing our numbers, this one accords best with their crusade to disseminate technologies that improve human health, promote human freedom, and conduce to human happiness.

Moving along from Priestley to his Victorian successor Herbert Spencer, recall that Spencer had his own answer to Malthus, in which he suggested that population increase was such a spur to human inventiveness that it would bring about its own resolution:

> Evidently, so long as the fertility of the race is more than sufficient to balance the diminution by deaths, population must continue to increase: so long as population continues to increase, there must be pressure on the means of subsistence: and so long as there is pressure on the means of subsistence, further mental development must go on, and further diminution of fertility must result. Hence, the change can never cease until the rate of multiplication is just equal to the rate of mortality; that is—can never cease until, on the average, each pair brings to maturity but two children.[7]

As always, Spencer's optimism was wildly overblown, but his projection of a relation between population pressure, economic development, and slowing birthrates has been borne out by the facts of declining fertility in all postindustrial economies. Ever the Lamarckian, Spencer thought that "mental development" was a biological inevitability tied to diminishing fertility, rather than a cultural achievement tied to family planning. He thought, more or less, that bigger brains equaled smaller wombs, and that the problem would take of itself. But despite the outmoded biology, he was pretty spot-on about the changing balance of life and death in the human species. Cultural evolution is Lamarckian, after all.[8] Moreover, even Spencer's *biological* Lamarckism doesn't seem quite so outlandish these days, in the light of our understanding of the way in which environmental stimuli control gene expression.

Spencer is the most unfashionable of Victorian intellectuals, routinely dismissed as a libertarian bigot who constructed an apologia for capitalism out of some nasty industrial ingredients, including "the laissez faire principles of the Anti-Corn Law League, the gloomy prognoses of Malthus, and the conservation of energy."[9] But Spencer was a social Lamarckian, in every way the opposite of the callous Malthusian of posthumous reputation. It was his belief in progress that caused him to rule out the randomness of Darwinian natural selection. For the same reason, he rejected the notion of human free will. Both were too capricious for a utopian of his stripe, threatening to throw a spanner into his mechanism of inexorable improvement.

Toward the end of his life, however, he arrived at a more sober assessment of our prospects for progress. During the Spanish-American War of 1898, a pacifist wrote to him with an appeal for help in forming a committee to arbitrate international conflict. Spencer's reply was prophetic: "There is a bad time coming," he wrote, "and civilized mankind will (morally) be uncivilized before civilization can again advance. . . . The universal aggressiveness and universal culture of blood-thirst will bring back military despotism, out of which after many generations partial freedom may again emerge."[10] In his eightieth year, Spencer looked down into the abyss of global war. Anticipating the scale of the catastrophe to come, he had to admit that humans must, on the evidence, be free to make suicidal mistakes. Finally, he accepted the possibility of regress.

Spencer foretold the apocalypse, and then tried to see beyond it. "After many generations, partial freedom may again emerge," he suggested to his correspondent. Are we getting any closer? The first half of the twentieth century showed without a shadow of doubt that we can go backward as well as forward. Progress itself was exposed as a problematic concept, tied to the arrogance of European colonizers and belied by their cruelty. But can we really do without a theory of progress? Surely insisting that nothing can get better is just as lopsided as insisting that everything will. Can we not dissect the aspirations of the Enlightenment a bit more delicately? Spencer might have dated the reemergence of his ideals of freedom to 1978, the year that saw an explosion in free markets, a sharp uptick in invocations of human rights on the world stage,[11] the drawing up of the uncannily Spencerian *Four Principles of Biomedical Ethics*, and the elimination of smallpox in accordance with Jenner's optimist forecast of 180 years earlier.

The jury is still out on whether neoliberalism constitutes progress or regress, but if there was one thing that all my British philosophers agreed upon, utilitarians and anti-utilitarians alike, it was that progress could be measured in hedonistic terms. "No school [of ethics]," Spencer declared, "can avoid taking for the ultimate moral aim a desirable state of feeling called by whatever name—gratification, enjoyment, happiness."[12] They are certainly in good company now. The last two decades have seen the rise of happiness and pleasure studies commensurate with the obsession that engulfed nineteenth-century Britain. And in the company of today's hedonist philosophies, Spencer's musings seem rather wise.

Writing against the grain of a more sternly moralizing society than our own, in which ideas of asceticism and sacrifice loomed so much larger, Spencer was a great proponent of the subversive notion that the cultivation of one's own joy was a moral imperative. Accordingly, his "four principles" of ethics included "self-care," the precept that "each individual shall perform all those acts required to fill up the measure of his own private happiness."[13] Much of *The Data of Ethics*, a book that Spencer dictated to his secretary from his sickbed when he thought he might be dying, is devoted to exploring the principle of self-care, which turns out to be the key to his whole moral philosophy.

For all he believed in personal happiness, Spencer knew that much of life is a trade-off. Most of us have to work at unpleasant jobs. Parents give up their freedom for their children. Neglect of these duties to self or other produces pain and chaos; such activities are therefore what he called "relatively right." All sacrificing of one's own comfort or happiness to avert a greater evil falls into this category. But there is a class of acts, he argued, that has the character of absolute rightness: those in which the giving of pleasure is itself pleasurable.

For Spencer, the paradigm of absolute rightness is breast-feeding. Obviously, not every mother enjoys having her nipples chewed; the point is simply that for those who do, breast-feeding constitutes a particularly symbiotic form of mutual pleasure. Another charming example that he offers is the benign give-and-take between strangers on city streets. "Someone who has slipped is saved from falling by a bystander: a hurt is prevented and satisfaction is felt by both." Or: "A fellow passenger is about to alight at the wrong station, and, warned against doing so, is saved from evil: each being as a consequence, gratified." Elevating these trivial acts to the apex of his morality, Spencer celebrates the civilized texture of everyday urban life. His most enviable practitioner of absolute rightness is "the artist of genius—poet, painter, or musician" who "obtains the means of living by acts that are directly pleasurable to him, while they yield, immediately or remotely, pleasures to others."[14]

Spencer lists just these three examples of absolute rightness, but others abound. Akin to Spencer's artist of genius is the scientist working on a cure. Relieving someone's distress with a sympathetic ear is surely one of life's great, if solemn, pleasures. And what about the infectious delight of making everyone around a table burst out laughing? Of all the species of absolute rightness that I enjoy, perhaps none fits the Spencerian bill better than my garden. I enjoy working on it; my neighbors enjoy walking past it. In it I grow milkweed for the caterpillars of the monarch butterfly, a species that has suffered from habitat destruction owing to industrial agricultural methods. The organization Monarch Watch sent me the first set of seeds. Every spring the butterflies float into the garden like airborne stained glass, to have sex on the wing, lay eggs on the milkweed leaves, and drink the flowers' nectar. Their jade and gold chrysalises look like jewels. Their eggs are iridescent seed pearls. Even the caterpillars, chewing and shitting all day, are handsome little beasts.

The Restoration of the Monarchy may be starting to work. The other day I woke up to a story on the radio about the recovery of monarch populations in their overwintering sites in Mexico, partly because of the efforts of citizen conservationists who have cultivated milkweed in gardens all along the butterfly's migration route through the United States. My misanthropic Malthusian side thinks that the bottomless human appetite for pleasure, comfort, and convenience is to blame for the environmental crisis. It is all too easy to see how rampant consumerism has delivered us to the brink of self-destruction. The libertarians have captured the economic argument, the hedonists have captured the psychological one, and we are hurtling toward extinction, taking half the planet with us. But if the only humane way round the problem of pleasure is through it, I sometimes dare to hope that our unquenchable drive for aesthetic delight contains within it the seeds of its own sustainability. A degraded environment eventually robs us of the most fundamental gratifications of sentience—sweet air, clean water, health, life, wild beauty. Surely we can learn to align human economic and political freedom with replenishment and stewardship of the nonhuman world? No heavy-handed, top-down, neo-Malthusian solution will cut it, but maybe, just maybe, we can harness the power of pleasure.

NOTES

Introduction

1. Thomas W. Lippman, "Saigon Yields; New Exchange of POWs Set"; Bob Woodward and Carl Bernstein, "FBI Chief Says Nixon's Aides Paid Segretti"; Hobart Rowen, "Burns Urges Quick Action on Currency"; Leroy F. Aarons, "U.S. Gives Indians until Sundown to Give Up"; *Washington Post*, March 8, 1973.

2. For a great survey of the 1973 collapse of American legitimacy, see Andreas Killen, *1973 Nervous Breakdown: Watergate, Warhol, and the Birth of Post-Sixties America* (New York: Bloomsbury, 2006).

3. *Quality of Health Care—Human Experimentation: Hearings before the Committee on Labor and Public Welfare, Subcommittee on Health*, 93rd Cong.1033–43 (1973) (testimony of Fred Gray, attorney; Charles Pollard, Tuskegee Study survivor; and Lester Scott, Tuskegee Study survivor).

4. See, for example, articles 10, 11, 19, and 32 of the United Nations Declaration on the Rights of Indigenous Peoples. So ubiquitous is informed consent in this area that indigenous rights activists and their supporters refer to "free, prior, and informed consent" by the acronym FPIC, pronounced "eff-pick." As for sexual negotiations, here is the definition of "affirmative consent" used on many American campuses for the purposes of legislating sexual misconduct: "Consent is informed. Consent is an affirmative, unambiguous, and conscious decision by each participant to engage in mutually agreed-upon sexual activity. Consent is voluntary. It must be given without coercion, force, threats, or intimidation. Consent means positive cooperation in the act or expression of intent to engage in the act pursuant to an exercise of free will. Consent is revocable. Consent to some form of sexual activity does not

imply consent to other forms of sexual activity. Consent to sexual activity on one occasion is not consent to engage in sexual activity on another occasion. A current or previous dating or sexual relationship, by itself, is not sufficient to constitute consent. Even in the context of a relationship, there must be mutual consent to engage in sexual activity. Consent must be ongoing throughout a sexual encounter and can be revoked at any time. Once consent is withdrawn, the sexual activity must stop immediately. Consent cannot be given when a person is incapacitated. A person cannot consent if s/he is unconscious or coming in and out of consciousness. A person cannot consent if s/he is under the threat of violence, bodily injury or other forms of coercion. A person cannot consent if his/her understanding of the act is affected by a physical or mental impairment." See, for example, the text of California Senate Bill 967, signed into law by Governor Jerry Brown in 2014.

5. Fred Gray, *The Tuskegee Syphilis Study: The Real Story and Beyond* (Montgomery, AL: New South Books, 1998), 54.

6. Given the complexity of the disease, the many contradictory interpretations of the archive, and the unreliability of the historical record, it seems impossible to assess how much physical harm was done to the subjects.

7. Allan M. Brandt, "Racism and Research: The Case of the Tuskegee Syphilis Study," *Hastings Center Report* 8, no. 6 (1978): 27.

8. Susan Reverby, *Examining Tuskegee: The Infamous Syphilis Study and Its Legacy* (Chapel Hill: University of North Carolina Press, 2009), 163.

9. Robert Cooke and Patricia El-Hinnawy, "Interview with Robert E. Cooke: Belmont Oral History Project" (Office of Human Research Protections, US Department of Health and Human Services, 2004), 6.

10. Richard A. Shweder, "Tuskegee Re-examined," *Spiked*, January 8, 2004, http://www.spiked-online.com/newsite/article/14972#.VzWcU4QrLIV.

11. Tom Beauchamp, the author of the document commonly known as the Belmont Report (discussed extensively in chapter 6), said in a 2004 interview: "I think that respect for persons, actually, is the most important principle. And for this reason: It becomes a kind of threshold condition. If you don't have adequate informed consent, either of a first-party or a third-party, you can't proceed with the research. So, it has a kind of priority position in terms of telling you what can be done and what can't be done. The next-most important principle, by far, is the principle of beneficence, and justice, then, is sort of a trailing third." Other members of the commission were also asked which principle they thought was the most important, and almost all cited "respect for persons," the Belmont Report language for informed consent or autonomy. Tom Beauchamp and Patricia El-Hinnawy, "Interview with Tom Lamar Beauchamp: Belmont Oral History Project" (Office of Human Research Protections, US Department of Health and Human Services, 2004), 10.

12. See Ruth Richardson, *Death, Dissection, and the Destitute* (London: Routledge & Kegan Paul, 1987). My intellectual debts to this book cannot be overstated: it was Richardson's extraordinary history of the Anatomy Act that got me thinking about utilitarian medicine and Jeremy Bentham in the first place.

13. Rachel Haliburton, *Autonomy and the Situated Self: A Challenge to Bioethics* (Lanham, MD: Lexington Books, 2013), 19.

14. Jeremy Bentham, *An Introduction to the Principles of Morals and Legislation*, ed. J. H. Burns and H. L. A. Hart (Oxford: Oxford University Press, 1996), 11.

15. *Quality of Health Care—Human Experimentation*, 363–68 (testimony of Robert Galbraith Heath, chair of the Department of Psychiatry and Neurology, Tulane University).

16. Ibid., 365.

17. Ibid., 369–73 (testimony of B. F. Skinner, Professor of Psychology, Harvard University).

18. Bill Rushton, "The Mysterious Experiments of Dr. Heath: In Which We Wonder Who Is Crazy and Who Is Sane," *Courier*, August 29–September 4, 1974, 7.

19. Lauri Laitinen, "Ethical Aspects of Psychiatric Surgery," in *Neurosurgical Treatment in Psychiatry, Pain, and Epilepsy*, ed. William Herbert Sweet et al. (Baltimore: University Park Press, 1977), 487.

20. Robert G. Heath, *Exploring the Mind-Brain Relationship* (Baton Rouge, LA: Moran Printing, 1996), was eventually self-published.

21. Personal communications from Michael Bernstein, provost of Tulane University, September 2009 to May 2010.

22. Thanks again to Rachel Ponce for this insight.

23. Thanks to Chad Valasek for bringing this to my attention.

24. R. M. Hare, *Moral Thinking: Its Levels, Method, and Point* (Oxford: Oxford University Press, 1981), 132.

25. Ibid., 139.

26. Ibid., 44.

Chapter 1

1. The story of informed consent has been well told elsewhere, although with markedly different emphases. In the 1970s the first denunciatory narratives of postwar human experimentation in America appeared, most notably Jay Katz, Alexander Morgan Capron, and Eleanor Swift Glass, *Experimentation with Human Beings: The Authority of the Investigator, Subject, Professions, and State in the Human Experimentation Process* (New York: Russell Sage Foundation, 1972), which contains a collection of primary sources as well as a legal and ethical overview. The apotheosis of this tradition is George J. Annas and Michael A. Grodin, *The Nazi Doctors and the Nuremberg Code: Human Rights in Human Experimentation* (New York: Oxford University Press, 1992), which takes informed consent as a philosophical absolute, condemns American research for its failure to live up to this basic moral standard, and makes invidious comparisons to Nazi research. David J. Rothman, *Strangers at the Bedside: A History of How Law and Bioethics Transformed Medical Decision Making* (New York: Basic Books, 1991), is a pioneering historical assessment, sensitive to the context of World War II but still accepting the centrality of informed consent and asking why American researchers did not practice it. In 1995 the most

important study of pre–World War II medical ethics appeared: Susan Lederer, *Subjected to Science: Human Experimentation in America before the Second World War* (Baltimore: Johns Hopkins University Press, 1995), provided a much-needed corrective to the view that consent was always ignored before the Nuremberg Trial. A turning point in post–World War II historiography came with the *Final Report of the Advisory Committee on Human Radiation Experiments* (New York: Oxford University Press, 1996), informed by Lederer's work, which stressed the extent to which informed consent was adhered to in the 1950s, especially in the military. Since that time more revisionist histories have appeared, arguing for a more nuanced approach to the preexisting ethical standards. Sydney A. Halpern, *Lesser Harms: The Morality of Risk in Medical Research* (Chicago: University of Chicago Press, 2004), is especially valuable as the only narrative that does not assume informed consent as a given moral standard and then puzzle over doctors' failure to conform to it. Laura Stark, *Behind Closed Doors: IRBs and the Making of Ethical Research* (Chicago: University of Chicago Press, 2012), contains the fruits of a fascinating trawl through the archives of the National Institute of Health Clinical Center, where the Institutional Review Board was established as the legal arbiter of consent in the 1950s. My account is indebted to all of these texts, while at the same time treating them as primary documents in the story of changing standards and attitudes.

2. Telford Taylor, *The Anatomy of the Nuremberg Trials: A Personal Memoir* (New York: Knopf, 1992), is a memoir of the International Military Tribunal case, the first of the proceedings, and the only one that was under the jurisdiction of all four occupying powers. Taylor was also the chief prosecutor in the medical case, but unfortunately his promised memoir about those particular proceedings was never forthcoming.

3. Ulf Schmidt, *Justice at Nuremberg: Leo Alexander and the Nazi Doctors' Trial* (New York: Palgrave Macmillan, 2004), 123–26 (quotation, 125).

4. Leonard Rossinger and Myron Davies, "Prison Malaria: Convicts Expose Themselves to Disease So Doctors Can Study It," *Life* 18, no. 23 (June 4, 1945): 46.

5. *Final Report of the Advisory Committee on Human Radiation Experiments*, 131–36.

6. Schmidt, *Justice at Nuremberg*, 136–37. See also Paul Weindling, "The Origins of Informed Consent: The International Scientific Commission on Medical War Crimes and the Nuremberg Code," *Bulletin of the History of Medicine* 75, no. 1 (2001): 37–71.

7. *Trials of War Criminals before the Nuernberg Military Tribunals under Control Council Law no. 10, Nuremberg, October 1946–April 1949* (Washington, DC: US Government Printing Office, 1949), 1:71.

8. E. R. Cunniffe, "Supplementary Report of the Judicial Council," *Journal of the American Medical Association* 132, no. 17 (December 28, 1946): 1090.

9. Jon M. Harkness, "Nuremberg and the Issue of Wartime Experiments on US Prisoners," *Journal of the American Medical Association* 276, no. 20 (1996): 1672–75.

10. *Trials of War Criminals*, 2:82–87.

11. Ibid., 2:92.

12. Erich Hans Halbach and Hans O. Luxenburger, "Experiments on Human

Beings as Viewed in World Literature" and "Supplement to the Report, Experimentation on Human Beings as Viewed in World Literature," Becker-Freyseng Document Book IV, documents nos. 60 and 60a (1947), in Harvard Law School Library, "Nuremberg Trials Project," http://nuremberg.law.harvard.edu/documents/229, http://nuremberg.law.harvard.edu/documents/230 (accessed December 14, 2016).

13. *Trials of War Criminals*, 2:92.

14. Ibid., 2:42.

15. Ibid., 2:181–82. The consent requirement in this more elaborate form seems to have been the work of one of the prosecution investigators, neurologist and psychologist Leo Alexander, who was more committed than Ivy to informed consent as expressing an ideal of human moral freedom and self-determination, rather than as a pragmatic instrument to maintain the legitimacy of human experimentation.

16. "Eine neuartige Heilbehandlung darf nur vorgenommen werden, nachdem die betreffende Person oder ihr gesetzlicher Vertreter auf Grund einer vorangegangenen zwecksprechenden Belehrung sich in unzweideutiger Weise mit der Vornahme einverstanden erklaert hat." Hans-Martin Sass, "Reichsrundschreiben 1931: Pre-Nuremberg German Regulations Concerning New Therapy and Human Experimentation," *Journal of Medicine and Philosophy* 8, no. 2 (1983): 99–112, 108.

17. *Trials of War Criminals*, 2:83.

18. See Jochen Vollmann and Rolf Winau, "Informed Consent in Human Experimentation before the Nuremberg Code," *BMJ* 313, no. 7070 (1996): 1445–47.

19. E. Wallace, 1946, "Memoranda Concerning Experiments on Children in Post-War Germany." Document Karl Brandt, no. K.B. 93, in Harvard Law School Library, "Nuremberg Trial Project," http://nuremberg.law.harvard.edu/documents/2746. This memo never found its way into the abridged version of the proceedings that was published by the American government in 1949, excluded, no doubt, on the grounds that it was too embarrassing.

20. R. A. McCance, "The Practice of Experimental Medicine," *Proceedings of the Royal Society of Medicine* 44 (1951): 194.

21. In *Lesser Harms*, the sociologist Sydney Halpern presents a wide survey of pre–informed consent research ethics, which conform in every respect to McCance's scheme. Within this widely shared morality, which she aptly dubs the "logic of lesser harms," Halpern identifies three implicit commitments: (1) to consent for healthy subjects, (2) to therapeutic potential for patients, and (3) to the epistemic and moral values of scientific method. This combination she rightly identifies as the "indigenous ethics" of biomedical researchers. I am much indebted to Halpern's insightful book, which shows how consent was just one of a set of practices designed to advance research objectives, rather than an absolute standard from which "unethical" researchers occasionally deviated. I think she may, however, underestimate the problems with the utilitarian calculus as applied to research on institutionalized people.

22. Jonathan D. Moreno and Susan E. Lederer, "Revising the History of Cold War Research Ethics," *Kennedy Institute of Ethics Journal* 6, no. 3 (1996): 223–37.

23. Stark, *Behind Closed Doors*.

24. *Final Report of the Advisory Committee on Human Radiation Experiments,* 1:83–139, 150–62.

25. Halpern, *Lesser Harms,* 120, observes of this period that "researchers emphasized expected gains for medical knowledge and substantially relaxed their observation of lesser-harm logic when it came to outcomes for subjects," a development she characterizes as "utilitarian." Rothman, *Strangers at the Bedside,* makes the same point: "Utilitarian justifications that had flourished under conditions of combat and conscription persisted, and principles of consent and voluntary participation were often disregarded" (51).

26. Rothman, *Strangers at the Bedside,* 30–31.

27. Vannevar Bush, *Science, the Endless Frontier: A Report to the President on a Program for Postwar Scientific Research* (Washington, DC: Government Printing Office, 1945), 13.

28. Halpern, *Lesser Harms,* 120; *Final Report of the Advisory Committee on Human Radiation Experiments,* 152; Stark, *Behind Closed Doors,* 95–97.

29. Susan Reverby, author of many invaluable books and articles on the Tuskegee Study, came across this evidence in the archives of the Public Health Service. See "Ethically Impossible: STD Research in Guatemala from 1946 to 1948" (Washington, DC: Presidential Commission for the Study of Bioethical Issues, 2011), http://bioethics.gov/node/654 (accessed October 14, 2014).

30. For example, A. M. Chesney and J. E. Kemp, "Studies in Experimental Syphilis: VI. On Variations in the Response of Treated Rabbits to Reinoculation and on Cryptogenetic Reinfection with Syphilis," *Journal of Experimental Medicine* 44, no. 5 (1926): 589–606.

31. Many of the primary documents from the study have been made publically available through National Archives website, including Cutler's "Final Syphilis Report," from which this quotation is taken. These make even more disturbing reading than the report of the bioethics commission. John C. Cutler, "Final Syphilis Report," 1948, Record Group 442: Records of the Centers for Disease Control and Prevention, 1921–2006, National Archives: 6, https://nara-media-001.s3.amazonaws.com/arcmedia/research/health/cdc-cutler-records/folder-1-syphilis-report.pdf.

32. Ibid., 8.

33. "Ethically Impossible," 41–42.

34. Cutler, "Final Syphilis Report," 33.

35. "Ethically Impossible," 33.

36. Ibid., 57.

37. Ibid., 74.

38. Letter from Roberto Robles Chinchilla, chief of the medical service at the penitentiary where the experiment began, in Cutler, "Final Syphilis Report," immediately after the contents page.

39. Cutler went on to enjoy a distinguished career in American venereal disease research, including a stint overseeing the Tuskegee Study of Untreated Syphilis. Susan Reverby, *Examining Tuskegee: The Infamous Syphilis Study and Its Legacy* (Chapel Hill: University of North Carolina Press, 2009), 144–46.

40. David M. Oshinsky, *Polio: An American Story* (Oxford: Oxford University Press, 2005), 151.

41. Ibid., 203.

42. The full story can be found in Henning Sjöström and Robert Nilsson, *Thalidomide and the Power of the Drug Companies* (Harmondsworth: Penguin, 1972).

43. The pediatrician was Widikund Lenz, the son of Nazi race theorist Fritz Lenz. One could see his confrontation with the erstwhile Nazi medical officer Heinrich Mückter as an example of the Auschwitz generation defending itself against its horrified children. The fact that Widikund Lenz lost in this exchange reveals how the Cold War imperative to establish West Germany as an economic powerhouse with a runaway pharmaceutical sector ran roughshod over the lessons of World War II. For Fritz Lenz, see Robert Proctor, *Racial Hygiene: Medicine under the Nazis* (Cambridge, MA: Harvard University Press, 1988).

44. Richard Harris, *The Real Voice* (New York: Macmillan, 1964), 221.

45. Ibid., 208.

46. Ibid., 224; emphasis added.

47. Drug Amendments of 1962, Pub. Law No. 87-781, 76 Stat. 783.

48. Rothman, *Strangers at the Bedside*, 87.

49. Rebecca Skloot, *The Immortal Life of Henrietta Lacks* (New York: Crown Publishers, 2010).

50. Katz, Capron, and Glass, *Experimentation with Human Beings*, 2.

51. Ibid., 11.

52. Ibid., 16.

53. Elinor Langer, "Human Experimentation: Cancer Studies at Sloan-Kettering Stir Public Debate on Medical Ethics," *Science* 143, no. 3606 (1964), 551.

54. Katz, Capron, and Glass, *Experimentation with Human Beings*, 49.

55. Ibid, 47.

56. John A. Osmundsen, "Physician Scores Tests on Humans: He Asserts Experiments Are Done without Consent," *New York Times*, March 24, 1965, 35.

57. Henry Beecher, "Ethics and Clinical Research," in *Ethics in Medicine: Historical Perspectives and Contemporary Concerns*, ed. Stanley Joel Reiser, Arthur J. Dyck, and William J. Curran (Cambridge, MA: MIT Press, 1977), 288–93.

58. Ibid., 288; emphasis in original.

59. J. A. von Felsinger, L. Lasagna, and H. K. Beecher, "The Response of Normal Men to Lysergic Acid Derivatives (di- and mono-ethyl amides)," *Journal of Clinical and Experimental Psychopathology* 17 (1956): 414–28.

60. Henry Beecher, "Experimentation in Man," *Journal of the American Medical Association* 169, no. 5 (1959): 463.

61. As an anesthesiologist, Beecher was also involved in a series of fascinating debates about pain and its relief that serve as a counterpoint to his issues with cost-benefit thinking in medical research. See Keith Wailoo, *Pain, a Political History* (Baltimore: Johns Hopkins University Press, 2014).

62. This issue has not gone away, of course. The question of how the "information" in informed consent should be conveyed in such a way as to promote the volun-

tariness of the subject's decision continues to vex ethicists, but it no longer counts as a reason to dodge the obligation altogether.

63. Rothman, *Strangers at the Bedside*, 91–92.

64. The best discussion of theologians' involvement in bioethics during this period and after is John Hyde Evans, *The History and Future of Bioethics: A Sociological View* (New York: Oxford University Press, 2012).

65. Paul Ramsey, *The Patient as Person: Explorations in Medical Ethics* (New Haven, CT: Yale University Press, 1970), 2; emphasis added.

66. Ibid., xiii.

67. Shana Alexander, "They Decide Who Lives, Who Dies," *Life* 53, no. 19 (November 9, 1962): 110.

68. David Sanders and Jesse Durkeminier, "Medical Advance and Legal Lag: Hemodialysis and Kidney Transplantation," *UCLA Law Review* 15, no. 2 (1968): 377.

69. Ramsey, *Patient as Person*, 259.

70. James Childress, "Who Shall Live When Not All Can Live," in Reiser, Dyck, and Curran, *Ethics in Medicine*, 623.

71. Reverby, *Examining Tuskegee*, 70–71.

72. Ibid., 75.

73. Ibid., 76–78.

74. Ibid., 79–81.

75. Jean Heller, "Syphilis Victims in U.S. Study Went Untreated for 40 Years," *New York Times*, July 26, 1972.

76. James T. Wooten, "Survivor of '32 Syphilis Study Recalls a Diagnosis," *New York Times*, July 27, 1972.

77. Jane E. Brody, "All in the Name of Science," *New York Times*, July 30, 1972.

78. Frazier Benya, "Biomedical Advances Confront Society: Congressional Hearings and the Development of Bioethics, 1960–1975" (PhD, University of Minnesota, 2012), 186, http://conservancy.umn.edu/bitstream/handle/11299/141246/Benya_umn_0130E_13259.pdf?sequence=1&isAllowed=y.

79. *Quality of Health Care—Human Experimentation*, 1033–43 (Senator Edward Kennedy's opening remarks), 2.

80. Ibid., 3.

81. Ibid. (statement of Anna Burgess), 77.

82. Ibid. (statement of Leonard Brooks), 78.

83. Ibid. (statement of Leodus Jones), 825–26.

84. Ibid. (prepared statement on "Experimentation on Prisoners in the Philadelphia County Prisons"), 830.

85. Ibid. (statement of Jessica Mitford), 796.

86. Ibid. (statement of Bernard Diamond), 841–42.

87. Ibid. (statement of Alexander Morgan Capron), 843.

88. Ibid. (statement of Bertram S. Brown), 338–39.

89. Ibid. (statement of Orlando J. Andy), 351.

90. Ibid. (statement of Peter R. Breggin), 358.

91. Ibid. (statement of Robert G. Heath), 363–64.

92. Ibid., 365.

93. Ibid., 366–67.

94. Judith Hooper and Dick Teresi, *The Three-Pound Universe* (New York: Macmillan, 1986), 155–57. The fact that these two lots of footage seem to have been the ones Heath showed everyone makes me suspect that they were by far the best demonstrations of the power of the technique, and that the archive at Tulane contains a lot of film of much less convincing results.

95. *Quality of Health Care—Human Experimentation* (statement of Robert G. Heath), 367–68.

96. Ibid. (statement of B. F. Skinner), 369.

97. Ibid., 370–71.

Chapter 2

1. Jeremy Bentham, *Bentham's Auto-Icon and Related Writings*, ed. James E. Crimmins (Bristol: Thoemmes Press, 2002).

2. My deep indebtedness to historians of science Steven Shapin and Simon Schaffer will be evident to everyone who knows their work. Their coauthored book *Leviathan and the Air Pump* (Princeton, NJ: Princeton University Press, 1985) and Shapin's *Social History of Truth: Civility and Science in Seventeenth-Century England* (Chicago: University of Chicago Press, 1994) are the source for much of my understanding of the politics of science arising from the English Civil War. Much of this and the next chapter were inspired by Schaffer's brilliant article about Bentham and Priestley, "States of Mind: Enlightenment and Natural Philosophy," in *The Languages of Psyche: Mind and Body in Enlightenment Thought: Clark Library Lectures, 1985–1986*, ed. George Sebastian Rousseau (Berkeley: University of California Press, 1990), 233–90.

3. The line is from a metrical Latin autobiography penned by Hobbes and quoted by John Aubrey, *Brief Lives, Chiefly of Contemporaries, Set Down by John Aubrey, between the Years 1669 and 1696*, ed. Andrew Clark (Oxford: Clarendon Press, 1898), 1:328; See also A. P. Martinich, *Hobbes: A Biography* (Cambridge: Cambridge University Press, 1999); Quentin Skinner, *Visions of Politics: Hobbes and Civil Science* (Cambridge: Cambridge University Press, 2002).

4. Aubrey, *Brief Lives*, 1:387.

5. Ibid., 1:328–30.

6. Ibid., 1:330–31.

7. Ibid., 1:347.

8. Ibid., 1:347.

9. Ibid., 1:331.

10. Ibid., 1:348–49.

11. Ibid., 1:349.

12. Thomas Hobbes, *The English Works of Thomas Hobbes*, ed. William Molesworth (London: John Bohn, 1840), 4:414.

13. Aubrey, *Brief Lives*, 1:367.

14. Howard Jones, *The Epicurean Tradition* (London: Routledge, 1989), 178–

79; Catherine Wilson, *Epicureanism at the Origins of Modernity* (Oxford: Oxford University Press, 2008), 2–3.

15. Wilson, *Epicureanism*, 142.

16. See Thomas Hobbes, *The Correspondence of Thomas Hobbes*, ed. Noel Malcolm (Oxford: Oxford University Press, 1994), 141, for our hero's pragmatic attitude toward this dangerous appointment.

17. Thomas Hobbes, *De Cive; or, The Citizen*, ed. Sterling Lamprecht (New York: Appleton-Century-Crofts, 1949), 49.

18. Thomas Bayly Howell, *Cobbett's Complete Collection of State Trials, and Proceedings for High Treason and Other Crimes and Misdemeanours, from the Earliest Period to the Present Time* (London: Hansard, 1809), 4:1141–42.

19. Martinich, *Hobbes*, 213; Skinner, *Visions of Politics*, 19.

20. Thomas Hobbes, *Leviathan*, ed. C. B. Macpherson (Harmondsworth: Penguin, 1968), 120–21.

21. Ibid., 227.

22. Ibid., 125.

23. Ibid., 186.

24. Skinner, *Visions of Politics*, 23.

25. Aubrey, *Brief Lives*, 1:340.

26. Samuel Mintz, *The Hunting of Leviathan: Seventeenth-Century Reactions to the Materialism and Moral Philosophy of Thomas Hobbes* (Cambridge: Cambridge University Press, 2010), 50–51.

27. Hobbes, *Leviathan*, 127. This sentiment does not represent the whole of Hobbes's account of "deliberation"—he had things to say about science and about self-control, both of which he understood to be uniquely human capacities—but it was the one that resonated through the psychological sciences discussed in this book.

28. John Locke, *Two Treatises of Government*, ed. Peter Laslett, 2nd ed. (Cambridge: Cambridge University Press, 1970), 17.

29. John Locke, *Two Tracts on Government*, ed. Philip Abrams (Cambridge: Cambridge University Press, 1967), 119–21.

30. J. R. Milton, "Locke's Life and Times," in *The Cambridge Companion to Locke*, ed. Vere Chappell (Cambridge: Cambridge University Press, 1997), 8.

31. John Locke, *An Essay Concerning Toleration and Other Writings on Law and Politics, 1667–1683*, ed. J. R. Milton and Philip Milton (Oxford: Clarendon Press, 2006), 274.

32. Locke, *Two Treatises of Government*, 350, 368.

33. Ibid., 430.

34. For an excellent analysis of the wider context of Locke's liberalism, especially that of settler colonialism in America, see James Tully, "Placing the Two Treatises," in *Political Discourse in Early Modern Britain*, ed. Nicholas Phillipson and Quentin Skinner (Cambridge: Cambridge University Press, 1993), 253–80.

35. John Locke, *An Essay Concerning Human Understanding*, ed. Peter Nidditch (Oxford: Oxford University Press, 1975), 67.

36. Ibid., 130.

37. D. D. Raphael, ed., *British Moralists, 1650–1800* (Oxford: Clarendon Press, 1969), 1:141.

38. Ibid., 1:153.

39. Ibid., 1:284.

40. L. A. Selby-Bigge, *British Moralists: Being Selections from Writers Principally of the Eighteenth Century* (Oxford: Clarendon Press, 1897), 1:110–17.

41. David Hume, *Philosophical Works* (Bristol: Thoemmes Press, 1996), 4:295.

42. Claude Helvétius, *De l'Esprit; or, Essays on the Mind, and its several Faculties*, trans. William Mudford (London: M. Jones, 1807), 249.

43. D. W. Smith, "Helvétius and the Problems of Utilitarianism," *Utilitas* 5, no. 2 (1993): 275–76.

44. Cesare Beccaria, *An Essay on Crimes and Punishments* (London: J. Almon, 1767), 2, 22, 26, 43.

45. John Rae, *Life of Adam Smith* (London: Macmillan, 1895), 13. See also the highly recommended Nicholas Phillipson, *Adam Smith: An Enlightened Life* (London: Penguin: 2010).

46. Adam Smith, *The Theory of Moral Sentiments*, ed. D. D. Raphael and A. L. Macfie (Oxford: Clarendon Press, 1976), 83.

47. D. D. Raphael, "Hume and Adam Smith on Justice and Utility," *Proceedings of the Aristotelian Society* 73 (1973): 95.

48. Smith, *Theory of Moral Sentiments*, 90–91.

49. David Hartley, *Observations on Man: His Frame, His Duties and His Expectations*, 2nd ed. (London: J. Johnson, 1791), 1:112. In his own century, it was the combination of scientific materialism and Christian soulfulness that made Hartley so appealing. Joseph Priestley, of whom more below, based his own moral and educational philosophy on Hartley's; Samuel Taylor Coleridge named his first son David Hartley Coleridge. In the nineteenth century, once it had been stripped of its theological embellishments, it would be hailed as the first stab at so-called association psychology.

50. Joseph Priestley, *Political Writings*, ed. Peter Miller (Cambridge: Cambridge University Press, 1993), 13.

51. Ibid., 45.

52. Ibid., 10.

53. No recent biography of Bentham exists, and so this account is pieced together from an early-twentieth-century biography, the autobiography contained in the Bowring edition of Bentham's collected works, and the introductions to the successive volumes of the correspondence. Charles Milner Atkinson, *Jeremy Bentham: His Life and Work* (London: Methuen & Co., 1905), 5–6.

54. Ibid., 9–10.

55. Ibid., 13–17.

56. Ibid., 19–20.

57. Ibid., 22–26.

58. https://www.ucl.ac.uk/Bentham-Project/tools/neologisms (accessed May 2, 2014).

59. Bentham attributed the paternity of the greatest happiness principle to a number of different men. The fact that he opens this particular work with a nod to scientific progress in the form of "the all vivifying and subtle element of the air, so recently analyzed and made known to us," suggests that Joseph Priestley was uppermost in his thoughts at the time of writing. Jeremy Bentham, *The Works of Jeremy Bentham*, ed. John Bowring (New York: Russell & Russell, 1962), 1:227.

60. Ibid., 1:237.

61. Ibid., 1:245, 289.

62. Atkinson, *Jeremy Bentham*, 23.

63. Bentham, *Works of Jeremy Bentham*, 1:272.

64. David Armitage, *The Declaration of Independence: A Global History* (Cambridge, MA: Harvard University Press, 2007), 173, 175, 176.

65. Atkinson, *Jeremy Bentham*, 47.

66. Samuel Romilly, *Memoirs of the Life of Sir Samuel Romilly, Written by Himself* (London: John Murray, 1841), 1:318.

67. Jeremy Bentham, *An Introduction to the Principles of Morals and Legislation*, ed. J. H. Burns and H. L. A. Hart (Oxford: Clarendon Press, 1996), 1, 11.

68. Ibid., 74.

69. Ibid., lxxxiii.

70. Ibid., 104–5.

71. Ibid., 52–54.

72. Atkinson, *Jeremy Bentham*, 101.

73. Jeremy Bentham, *Rights, Representation and Reform: Nonsense upon Stilts and Other Writings on the French Revolution*, ed. Philip Scholfield, Catherine Pease-Watkin, and Cyprian Blamires (Oxford: Clarendon Press, 2002), 330. Philip Schofield, *Utility and Democracy: The Political Thought of Jeremy Bentham* (Oxford: Oxford University Press, 2006), contains a brilliant analysis of Bentham's aversion to "fictitious entities" that is especially helpful in regard to the doctrine of natural rights.

74. Jeremy Bentham, *The Panopticon Writings*, ed. Miran Božovič (London: Verso, 1995).

75. Jeremy Bentham, *The Correspondence of Jeremy Bentham*, vol. 4, *October 1788 to December 1793*, ed. Alexander Taylor Milne (London: Athlone Press, 1981), 232.

76. Ibid., 4:286.

77. Ibid., 4:359.

78. Ibid., 4:395.

79. Ibid., 4:450.

80. Schofield, *Utility and Democracy*, 110–11.

81. Gertrude Himmelfarb, "Bentham's Utopia: The National Charity Company," *Journal of British Studies* 10, no. 1 (November 1970): 80–125. Himmelfarb excavated Bentham's manuscript notes on the scheme in order to deliver a verdict on it very much in keeping with the general odium in which utilitarianism was held in the 1970s. It has to be admitted that her research shows Bentham in a poor light: one

note has him suggesting, for instance, that the Houses of Industry inmates could start working at the age of four, and lamenting the waste involved in the standard age of fourteen: "Ten precious years in which nothing is done! nothing for industry! nothing for moral and intellectual improvement!" (101).

82. Bentham, *Works of Jeremy Bentham*, 8:369–439.

83. Ibid., 8:381.

84. Ibid., 8:382, 384.

85. Janet Semple, *Bentham's Prison: A Study of the Panopticon Penitentiary* (Oxford: Clarendon Press, 1993), 218–53.

Chapter 3

1. M. J. Daunton, *Progress and Poverty: An Economic and Social History of Britain, 1700–1850* (Oxford: Oxford University Press, 1995).

2. Robert Potter, *Observation on the Poor Laws, on the Present State of the Poor, and Houses of Industry* (London: J. Wilkie, 1775), 31.

3. Adam Smith, *The Wealth of Nations*, books 1–3, ed. Andrew Skinner (London: Penguin, 1986), 182–83.

4. Jeremy Bentham, *The Correspondence of Jeremy Bentham*, vol. 3, *January 1781 to October 1788*, ed. Ian Christie (London: Athlone Press, 1971), 57.

5. Ibid., 3:68.

6. Joseph Townsend, *A Dissertation on the Poor Laws by a Well-Wisher to Mankind* (Berkeley: University of California Press, 1971), 27.

7. Ibid., 37–38.

8. William Paley, *Principles of Moral and Political Philosophy* (Boston: West and Richardson, 1815), 84–87.

9. T. R. Malthus, *The Works of Thomas Robert Malthus*, ed. E. A. Wrigley and David Souden (London: William Pickering, 1986), 1:8.

10. Ibid., 1:51–52.

11. Ibid., 3:487.

12. T. R. Malthus, *An Essay on the Principle of Population; or, A View of Its Past and Present Effects on Human Happiness; with an Inquiry into our Prospects Respecting the Future Removal or Mitigation of the Evils Which It Occasions* (London: J. Johnson, 1803), 493. The collected works does not contain the 1803 edition, which is the most explicitly utilitarian version, as indicated by the new subtitle.

13. Malthus, *Works of Thomas Robert Malthus*, 1:9.

14. Malthus, *Essay on the Principle of Population*, 531.

15. Raymond G. Cowherd, *Political Economists and the English Poor Laws: A Historical Study of the Influence of Classical Economics on the Formation of Social Welfare Policy* (Athens: Ohio University Press, 1977), 57.

16. Ibid., 49–81.

17. "Review of Report on the Poor Laws," *Christian Observer*, January 1818, 42.

18. Cowherd, *Political Economists*, 133.

19. Patricia James, *Population Malthus: His Life and Times* (London: Routledge & Kegan Paul, 1979), 115.

20. Thomas Beddoes, *Contributions to Physical and Medical Knowledge, Principally from the West of England* (London: T. M. Longman and O. Rees, 1799), 147.

21. Malthus, *Works of Thomas Robert Malthus*, 1:65.

22. For the whole story, delightfully told, see Mike Jay, *The Atmosphere of Heaven: The Unnatural Experiments of Dr. Beddoes and His Sons of Genius* (New Haven, CT: Yale University Press, 2009). Jay's account of Davy's early years is on page 151.

23. Ibid., 172.

24. Humphry Davy, *Researches, Chemical and Philosophical; Chiefly Concerning Nitrous Oxide or Dephlogisticated Nitrous Air, and Its Respiration* (London: J. Johnson, 1800), 487–89.

25. Thomas Wedgwood, *The Value of a Maimed Life: Extracts from the Manuscript Notes of Thomas Wedgwood*, ed. Margaret Olivia Tremayne (London: C. W. Daniel, 1912), 39, 76, 73.

26. Davy, *Researches*, 519.

27. Jeremy Bentham, *The Works of Jeremy Bentham*, ed. John Bowring (New York: Russell & Russell, 1962), 1:300; Charles Milner Atkinson, *Jeremy Bentham: His Life and Work* (London: Methuen & Co., 1905), 151.

28. Bentham, *Works of Jeremy Bentham*, 1:304–5.

29. Charles Bell, *Letters of Sir Charles Bell, Selected from his Correspondence with his Brother George Joseph Bell* (London: John Murray, 1870), 275–76.

30. Bentham, *Works of Jeremy Bentham*, 1:197–205.

31. Alexander Bain, *James Mill: A Biography* (London: Longmans, Green & Co., 1882).

32. Ibid., 135.

33. James Mill, *The History of British India* (London: James Madden & Co., 1840), 2:150.

34. Jeremy Bentham, *Plan of Parliamentary Reform, in the form of a Catechism* (London: R. Hunter, 1817), 9.

35. James Mill, *Analysis of the Phenomena of the Human Mind* (London: Longmans, 1869), 2:185.

36. Ibid., 2:188.

37. In 1828 the devout utilitarian William Bentinck became governor general of India and left for the subcontinent promising to build the Panopticon in Calcutta. See Bentham to Col. Young, December 28, 1827, in Bentham, *Works of Jeremy Bentham*, 10:576–78.

38. Eric Stokes, *The English Utilitarians and India* (Oxford: Clarendon Press, 1963), 70.

39. Bart Schultz and Georgios Varouxakis, eds., *Utilitarianism and Empire* (Lanham, MD: Lexington Books, 2005). See especially the first essay, "Bentham on Slavery and the Slave Trade," by Frederick Rosen, which persuasively lays the responsibility for the heavy-handed utilitarian presence in India at Mill's door rather than Bentham's.

40. As one psychoanalytically inclined historian has noted, John Stuart Mill's

Autobiography reads as if he were "brought up without a mother." Bruce Mazlish, *James and John Stuart Mill: Father and Son in the Nineteenth Century*, 2nd ed. (New Brunswick, NJ: Transaction Inc., 1988), 152. Harriet Burrow, James Mill's wife and John Stuart's mother, was brought up in a private asylum in Hoxton owned by her family. The marriage seems to have been a disappointment to both parties and was widely known to be unhappy.

41. John Stuart Mill, *Collected Works of John Stuart Mill*, ed. John M. Robson and Jack Stillinger (London: Routledge, 1981), 1:9.

42. Nicholas Capaldi, *John Stuart Mill: A Biography* (Cambridge: Cambridge University Press, 2004), 19–20.

43. Mill, *Collected Works*, 1:81.

44. Ibid., 1:83.

45. Ibid., 1:143.

46. Ibid., 1:141, 139.

47. Ibid., 1:145. Others have engaged in retrospective diagnoses of Mill's bout of depression and dismissed his own characterization of the episode as pre-Freudian naiveté. Given that Mill attributed the beginning of his recovery to reading in a well-known French memoir about the death of the author's father, a Freudian diagnosis is well nigh irresistible. There seems to me no contradiction between this and the more general sense of oppression that Mill felt about his education at his father's hands. Harold Bloom's helpful formulation "anxieties of influence" nicely captures these two dimensions of the Oedipus complex. See Mazlish, *James and John Stuart Mill*, and Michele Green, "Sympathy and Self-Interest: The Crisis in Mill's Mental History," *Utilitas* 1 (1989): 259–77. Leslie Stephen makes a most plausible diagnosis of over-work, as a result of editing a huge manuscript of Bentham's *Rationale of Evidence* and superintending the five large volumes through the press. Leslie Stephen, *The English Utilitarians* (London: Duckworth and Co., 1900), 19.

48. Mill, *Collected Works*, 1:143.

49. Ibid., 1:141.

50. Ibid., 1:111.

51. Ibid., 18:223.

52. Ibid., 18:226.

53. Ibid., 10:212.

54. Charles Dickens, *Hard Times* (London: Wordsworth Classics, 1995), 78.

55. Mill, *Collected Works*, 1:107.

56. Richard Reeves, *John Stuart Mill: Victorian Firebrand* (New York: Atlantic, 2008), 1–2.

57. Mill, *Collected Works*, 26:286.

58. Ibid., 26:288.

59. Ibid., 4:375.

60. The definitive history of the Anatomy Act, and, indeed, of utilitarian medicine in the nineteenth century, is Ruth Richardson, *Death, Dissection, and the Destitute* (London: Routledge & Kegan Paul, 1987).

61. This sentiment from an otherwise sympathetic reviewer of Mackenzie's

pamphlet is representative: "There are thousands who would perish in their garrets and cellars, rather than go to an infirmary or workhouse, when they know that, in case of their death . . . they were to be hacked to pieces by the surgeon." "Review of William Mackenzie 'An Appeal to the Public and to the Legislature, on the Necessity of Affording Dead Bodies to the Schools of Anatomy, by Legislative Enactment,'" *Literary Chronicle and Weekly Review*, no. 298 (January 29, 1825), 70.

62. Thomas Southwood Smith, *Use of the Dead to the Living: From the Westminster Review* (Albany: Webster and Skinners, 1827), 37. The article first appeared in the *Westminster Review* in July 1824 but was reprinted in pamphlet form and went through a number of editions.

63. "Report from the Select Committee Appointed by the House of Commons to Inquire into the Manner of Obtaining Subjects for Dissection in the Schools of Anatomy, &c.," in *Papers Relative to the Question of Providing Adequate Means for the Study of Anatomy* (London: C. Adlard, 1832), 21–22.

64. Richardson, *Death, Dissection, and the Destitute*, 131–46.

65. C. L. Lewes, *Dr Southwood Smith: A Retrospect* (Edinburgh: William Blackwood and Sons, 1898), 46–47.

66. Richardson, *Death, Dissection, and the Destitute*, 184.

67. Ibid., 178.

68. Ibid., 228–30.

69. "Riot at Swan Street Cholera Hospital," *Manchester Guardian*, September 9, 1832, 3.

70. Benjamin Ward Richardson, *The Health of Nations: A Review of the Works of Edwin Chadwick, with a Biographical Dissertation* (London: Longmans, Green & Co., 1887), 127.

71. George Nicholls, *A History of the English Poor Law, in Connection with the State of the Country and the Condition of the People*, 2nd ed. (New York: G. P. Putnam and Sons, 1898), 2:301–2.

72. David Roberts, "How Cruel Was the Victorian Poor Law?" *Historical Journal* 6, no. 1 (1963): 106, 103, 105.

Chapter 4

1. Alexander Bain, *Autobiography* (London: Longmans, Green & Co., 1904), 15.

2. Ibid., 2–5.

3. Ibid., 14.

4. Ibid., 15–16, 24–25.

5. Ibid., 20–21.

6. Ibid., 49.

7. Alexander Bain, *John Stuart Mill: A Criticism with Personal Reflections* (London: Longmans, Green & Co., 1882), 64.

8. Alexander Bain, *The Senses and the Intellect* (London: Longmans, Green & Co., 1868), 8.

9. Ibid., 286.

10. Alexander Bain, *The Emotions and the Will* (London: Longmans, Green & Co., 1875), 274–75.

11. Ibid., 278.

12. Herbert Spencer, *Autobiography* (London: Williams and Norgate, 1904), 1:260.

13. Herbert Spencer, *Social Statics; or, The Conditions Essential to Human Happiness Specified, and the First of them Developed* (London: John Chapman, 1851), 8–10. It should be noted, *contra* his reputation for extreme conservatism, that Spencer's antipathy to all kinds of poor relief—direct payments and workhouses alike—was based on his opposition to private property in land, which he considered the fountainhead of all economic injustice.

14. Spencer was stingy with praise for any other ethical system, and liked to boast that he never read books (judging from his writings, this claim does not seem too exaggerated), but Adam Smith did get a grudging acknowledgement in *Social Statics* for having anticipated aspects of Spencer's scheme in his *Theory of Moral Sentiments*; Spencer, *Social Statics*, 96–97.

15. Ibid., 68–69.

16. Ibid., 78.

17. Ibid., 14.

18. Herbert Spencer, "A Theory of Population, Deduced from the General Law of Animal Fertility," *Westminster Review* 57, no. 112 (1852): 267.

19. Herbert Spencer, *Principles of Psychology* (London: Williams and Norgate, 1870), 490. See David Weinstein, *Equal Freedom and Utility: Herbert Spencer's Liberal Utilitarianism* (Cambridge: Cambridge University Press, 1998), esp. chap. 2.

20. Herbert Spencer, "Bain on the Emotions and the Will," *Medico-Chirurgical Review*, January1860, 65.

21. Bain, *Emotions and the Will*, 49.

22. Spencer, *Autobiography*, 1:15–68.

23. Bain, *Emotions and the Will*, 49–50.

24. Herbert Spencer, *The Data of Ethics* (London: G. Norman and Son, 1879).

25. Spencer, *Social Statics*, 183.

26. Barbara Taylor, *The Last Asylum: A Memoir of Madness in Our Times* (Chicago: University of Chicago Press, 2015), 112.

27. For the context within which West Riding was built, see Janet Oppenheim, *"Shattered Nerves": Doctors, Patients, and Depression in Victorian England* (New York: Oxford University Press, 1991); Andrew T. Scull, *The Most Solitary of Afflictions: Madness and Society in Britain, 1700–1900* (New Haven, CT: Yale University Press, 1993); Andrew T. Scull, Charlotte MacKenzie, and Nicholas Hervey, *Masters of Bedlam: The Transformation of the Mad-Doctoring Trade* (Princeton, NJ: Princeton University Press, 1996).

28. Samuel Tuke, *Practical Hints on the Construction and Economy of Pauper Lunatic Asylums, including Instructions to the Architects who offered Plans for the Wakefield Asylum* (York: William Alexander, 1815), 37.

29. Ibid., 28.

30. Samuel Tuke, "Plans, Elevations, Sections and Descriptions of the Pauper Lunatic Asylum Lately Erected at Wakefield by Messrs. Watson and Pritchett, Including Instructions Drawn Up by Mr. Samuel Tuke for the Architects Who Prepared Designs for the West Riding Asylum" (1819), unpaginated, West Yorkshire Archive Service, Wakefield.

31. As the architectural historian Robin Evans has noted, "the logic of inspection . . . remained from 1800 onwards the predominant organizing force in prison architecture." Robin Evans, *The Fabrication of Virtue: English Prison Architecture, 1750–1840* (Cambridge: Cambridge University Press, 1982), 276.

32. Jeremy Bentham, *The Panopticon Writings*, ed. Miran Božovič (London: Verso, 1995), 82.

33. Tuke, "Plans, Elevations, Sections and Descriptions."

34. Tuke, *Practical Hints*, 33.

35. William Charles Ellis, *A Treatise on the Nature, Symptoms, Causes, and Treatment of Insanity* (London: Samuel Holdsworth, 1838), 6.

36. William Charles Ellis and Charles Corsellis, "West Riding Lunatic Asylum Annual Reports, 1819–1847," 97, West Yorkshire Archive Service, Wakefield.

37. Ibid., 103.

38. Ibid., 111.

39. Ibid., 105.

40. Ibid., 103.

41. Ibid., 113.

42. Ibid., 120.

43. James Crichton Browne, "Medical Director's Journal: Reports to the Quarterly Meetings, 1867–1874," April 29, 1869, West Yorkshire Archive Service, Wakefield.

44. Ibid., July 30, 1874.

45. James Crichton-Browne, *Report of the Committee of Visitors and of the Medical Superintendent of the West Riding Pauper Lunatic Asylum for the Year 1868* (Wakefield: Hicks and Allen, 1869), 27.

46. For the Turkish baths, see James Crichton-Browne, *Report of the Committee of Visitors and of the Medical Superintendent of the West Riding Pauper Lunatic Asylum, for the Year 1871* (Wakefield: B. W. Allen, 1872), 26–28. For the play, see "The Drama in Lunacy Treatment," *Medical Times and Gazette*, April 17, 1875.

47. For a summary of his efforts in the cause of basic science, see Crichton-Browne, *Report . . . for the Year 1868*, 25–30.

48. James Crichton-Browne, "Cranial Injuries and Mental Diseases" *West Riding Lunatic Asylum Medical Reports*, ed. J. Crichton-Browne (London: J. & A. Churchill, 1871), 1:23–26.

49. James Crichton-Browne, *Report of the Committee of Visitors and of the Medical Superintendent of the West Riding Pauper Lunatic Asylum for the Year 1871*, 27–28.

50. For the nineteenth- and early-twentieth-century antivivisection movement

and its relation to medical ethics, see Susan E. Lederer, *Subjected to Science: Human Experimentation in America before the Second World War* (Baltimore: Johns Hopkins University Press, 1995). Lederer shows that it was the campaigners against vivisection who drew attention to the issue of research on human subjects, which they regarded as a natural outgrowth of the desensitizing effects of animal experiments.

51. Macdonald Critchley and Eileen A. Critchley, *John Hughlings Jackson: Father of English Neurology* (New York: Oxford University Press, 1998).

52. See John Hughlings Jackson, *Selected Writings of John Hughlings Jackson: On Epilepsy and Epileptiform Convulsions* (London: Hodder and Stoughton, 1931).

53. G. Fritsch and E. Hitzig, "On the Electrical Excitability of the Cerebrum" (1870), in Gerhardt von Bonin, *Some Papers on the Cerebral Cortex, Translated from the French and German* (Springfield, IL: Thomas, 1960), 73–96.

54. Alastair Compston, "Editorial," *Brain* 134 (2011): 1251–53.

55. David Ferrier, "Experimental Researches in Cerebral Physiology and Pathology," *West Riding Lunatic Asylum Medical Reports*, ed. J. Crichton Browne (London, J & A Churchill, 1873), 3:30–96.

56. David Ferrier, "Experiments on the Brains of Monkeys: First Series," *Proceedings of the Royal Society* 23 (1874–1875): 410.

57. Roberts Bartholow, "Experimental Investigations into the Functions of the Human Brain," *American Journal of the Medical Sciences* 134 (April 1874): 309.

58. Ibid., 310–11.

59. Ibid., 311.

60. David Ferrier, "Bartholow's Experiments on the Human Brain," *London Medical Record* 2 (1874): 285–86.

61. Lauren Julius Harris and Jason B. Almerigi, "Probing the Human Brain with Stimulating Electrodes: The Story of Roberts Bartholow's (1874) Experiment on Mary Rafferty," *Brain and Cognition* 70, no. 1 (2009): 92–115.

62. David Ferrier, "Experiments on the Brains of Monkeys: Second Series," *Philosophical Transactions of the Royal Society of London* 165 (1875): 433–88.

63. David Ferrier, *The Functions of the Brain* (London: Smith, Elder & Co., 1876), 288.

64. Ibid., 256; emphasis in original.

65. Ibid., 281.

66. Ibid., 296–97.

67. Ibid., 305.

68. Alex Saluka, "Baroness Burdett-Coutt's Garden Party: The International Medical Congress, London, 1881," *Medical History* 26, no. 2 (1982): 183–90.

69. "The Festivities of the Congress," *British Medical Journal* 2, no. 1076 (August 13, 1881): 303–4.

70. Ibid.

71. Bruno Latour, *The Pasteurization of France* (Cambridge, MA: Harvard University Press, 1988).

72. The leader of the antivivisection movement in Bismarckian Germany was an explorer by the name of Baron Ernst von Weber, who had embarked on a successful

agitprop career upon his return from a four-year trip to Africa. He published a volume called *The Torture-Chambers of Science* (*Die Folterkammern der Wissenschaft*), in which he narrated the details of various experiments lifted from the pages of physiology journals such as *Pflügers Archiv*. Von Weber was enraged by Goltz's boast that he was better than any of his peers at keeping animals alive with really extensive experimental lesions, and disgusted at the jocular tone with which the physiologist reported the behavioral deficits of his victims. Goltz hit back, calling *The Torture-chambers of Science* "a book of fairy-tales" and "a mish-mash of irrationality and malice." He also pointed out that von Weber was silent on the fact that Goltz anaesthetized his animals with chloroform before operating on them, which was clearly stated in the published reports of his experiments. Von Weber responded that the heinous act was not so much the surgical intervention as the continued study of an animal afterward, the same accusation that would later be leveled against Ferrier in court. See Hubert Bretschneider, *Der Streit um die Vivisektion im 19. Jahrhundert* (Stuttgart: Gustav Fischer, 1962), 40–44.

73. *Transactions of the International Medical Conference Seventh Session* (London: Ballantyne Hanson, 1881), 218–28.

74. Ibid., 228–33.

75. Ibid., 234–40.

76. Ibid., 237.

77. "The Charge against Professor Ferrier under the Vivisection Act: Dismissal of the Summons," *British Medical Journal* 2, no. 1090 (November 19, 1881): 836–42.

78. "Dr. Ferrier's Localizations: For Whose Advantage?" *British Medical Journal* 2, no. 1090 (November 19, 1881): 823.

79. Ibid.

80. Stanley Finger, *Minds behind the Brain* (Oxford: Oxford University Press, 2000), 173–74.

81. Kenneth Tyler and Rolf Malessa, "The Goltz-Ferrier Debates and the Triumph of Cerebral Localizationist Theory," *Neurology* 55, no. 7 (2000): 1015. See also John D. Spillane, *The Doctrine of the Nerves* (Oxford: Oxford University Press, 1981), 394–400; Finger, *Minds behind the Brain*, 172–75; Jürgen Thorwald and Leon Banov, *The Triumph of Surgery* (New York: Pantheon, 1960), 96; Charles A. Ballance, *A Glimpse into the History of Surgery of the Brain* (London: Macmillan and Co., 1922), 84–90. A notable exception to this triumphalist consensus is provided by sociologist of science Susan Leigh Star, who perhaps leans too far in the other direction, dismissing the claim of a link between Ferrier's localization of brain function and successful neurosurgical interventions as "nonsense." Susan Leigh Star, *Regions of the Mind: Brain Research and the Quest for Scientific Certainty* (Stanford, CA: Stanford University Press, 1989), 58.

82. Jeremy Bentham, *An Introduction to the Principles of Morals and Legislation*, ed. J. H. Burns and H. L. A. Hart (Oxford: Clarendon Press, 1996), chap. 17, n. 122.

83. Peter Vorzimmer, "Darwin, Malthus, and the Theory of Natural Selection,"

Journal of the History of Ideas 30, no. 4 (October–December 1969), claims that Darwin read the sixth edition of 1826, substantially similar to the 1803 edition.

84. Robert Richards, *Darwin and the Emergence of Evolutionary Ideas of Mind and Behavior* (Chicago: University of Chicago Press, 1987), 120.

85. Charles Darwin, *The Descent of Man and Selection in Relation to Sex* (New York: D. Appleton and Company, 1871), 94.

86. Charles Darwin, *The Expression of the Emotions in Man and Animals* (New York: D. Appleton and Company, 1886), 9.

87. Ibid., 215.

88. Charles Darwin, *The Life and Letters of Charles Darwin, Including an Auto-biographical Chapter*, ed. Francis Darwin (London: John Murray, 1888), 1:310.

89. Conwy Lloyd Morgan, *Animal Life and Intelligence* (Boston: Ginn and Co., 1891), 380.

90. Conwy Lloyd Morgan, *Habit and Instinct* (London: E. Arnold, 1896), 149.

Chapter 5

1. Conwy Lloyd Morgan, *Habit and Instinct* (London: E. Arnold, 1896), 35, 34, 82, 63, 85.

2. Conwy Lloyd Morgan, *An Introduction to Comparative Psychology* (London: Walter Scott, 1894), 53.

3. Geraldine M. Jonçich, *The Sane Positivist a Biography of Edward L. Thorndike* (Middletown, CT: Wesleyan University Press, 1968), 142.

4. Ibid., 126–42.

5. Edward Lee Thorndike, *Animal Intelligence: An Experimental Study of the Associative Processes in Animals* (New York: Macmillan, 1898), 3.

6. Edward Lee Thorndike, *Animal Intelligence: Experimental Studies* (New York: Macmillan, 1911), 74.

7. Thorndike, *Animal Intelligence: An Experimental Study*, 13.

8. Thorndike, *Animal Intelligence: Experimental Studies*, 244.

9. Robert Boakes, *From Darwin to Behaviorism: Psychology and the Minds of Animals* (Cambridge: Cambridge University Press, 1984), 13. This usage seems to have originated with James Mark Baldwin, *Mental Development in the Child and the Race: Methods and Processes* (New York: Macmillan and Co., 1895), 181. Boakes asserts that it was widely used until 1911.

10. Hurley Cason, "The Pleasure-Pain Theory of Learning," *Psychological Review* 39, no. 5 (1932): 440–66.

11. R. Woodworth, "Edward Lee Thorndike, 1874–1949," *Biographical Memoirs of the National Academy of Sciences of the United States of America* 27 (1952): 209–37.

12. E. L. Thorndike, "Science and Values," *Science* 83, no. 2140 (January 3, 1936): 1–2.

13. Kerry W. Buckley, *Mechanical Man: John Broadus Watson and the Beginnings of Behaviorism* (New York: Guilford Press), 7.

14. Ibid., 15.

15. John B. Watson, "Psychology as the Behaviorist Views It," *Psychological Review* 20, no. 2 (1913): 158.

16. Ibid.

17. Buckley, *Mechanical Man*, 153.

18. Bertrand Russell, *An Outline of Philosophy* (London: Routledge, 1993), 24.

19. Bertrand Russell, *The Autobiography of Bertrand Russell*, vol. 1, *1872–1914* (New York: Bantam Books, 1967), 8, 10.

20. Russell, *Outline of Philosophy*, 23.

21. B. F. Skinner, *The Shaping of a Behaviorist* (New York: Alfred A. Knopf, 1979), 31, 87.

22. Ibid., 61.

23. Ibid., 97, 143.

24. B. F. Skinner, *The Behavior of Organisms: An Experimental Analysis* (New York: Appleton-Century-Crofts, 1938), 22.

25. Skinner, *Shaping of a Behaviorist*, 233.

26. Ibid., 95.

27. C. B. Ferster and B. F. Skinner, *Schedules of Reinforcement* (East Norwalk, CT: Appleton-Century-Crofts, 1957), chap. 7.

28. Skinner, *Shaping of a Behaviorist*, 113.

29. Ibid., 241.

30. James Capshew, "Engineering Behavior: Project Pigeon, World War II, and the Conditioning of B. F. Skinner," in *B. F. Skinner and Behaviorism in America Culture*, ed. Laurence Smith and William Woodward (London: Associated University Presses, 1996), 128–50.

31. B. F. Skinner, *Walden Two* (New York: Macmillan, 1970), 76.

32. Hilke Kuhlmann, *Living Walden Two: B. F. Skinner's Behaviorist Utopia and Experimental Communities* (Urbana: University of Illinois Press, 2005), 8.

33. Skinner, *Walden Two*, 249.

34. Ibid., 296.

35. Ibid., 264.

36. B. F. Skinner, *A Matter of Consequences* (New York: Alfred Knopf, 1983), 25–26.

37. "It is a great deal easier to assume that pleasure stamps in the successful movement and that displeasure stamps out the unsuccessful and to let the matter rest than to institute the necessary experimentation." John B. Watson, *Behavior: An Introduction to Comparative Psychology* (New York: H. Holt and Co., 1914), 276.

38. J. Olds and P. Milner, "Positive Reinforcement Produced by Electrical Stimulation of the Septal Area and Other Regions of Rat Brain," *Journal of Comparative and Physiological Psychology* 47, no. 6 (1954): 426.

39. Intellectual historian Daniel Rogers has argued that utilitarianism had never got much rhetorical purchase in America precisely because of the deep cultural commitment to the ideal of natural rights upon which the nation had been founded. Daniel T. Rodgers, *Contested Truths: Keywords in American Politics since Indepen-*

dence (New York: Basic Books, 1987), 17–79. See especially the George Bernard Shaw quotation, 37.

40. Jamie Cohen-Cole, *The Open Mind: Cold War Politics and the Sciences of Human Nature* (Chicago: University of Chicago Press, 2014).

41. Theodor W. Adorno, *The Authoritarian Personality* (New York: Harper, 1950), x.

42. Ibid., 976.

43. Ibid., 759.

44. Cohen-Cole, *Open Mind*, 43.

45. Daniel Brower, "The Problem of Quantification in Psychological Science," *Psychological Review* 56, no. 6 (1949): 328, 332.

46. Cohen-Cole, *Open Mind*, 85.

47. Ibid., 149.

48. Ibid., 169–74.

49. Wallace Tomlinson, "Interview with Robert Galbraith Heath," Tulane Medical Center, March 5, 1986, https://archive.org/details/WallaceTomlinsonInterviewing RobertHeath_March51986.

50. Sandor Rado, *Adaptational Psychodynamics*, ed. Jean Jameson and Henriette Klein (Northvale, NJ: Jason Aronson, 1995), 57.

51. Tomlinson, "Interview with Robert Galbraith Heath."

52. Fred Mettler, ed., *Selective Partial Ablation of the Frontal Cortex* (New York: Paul B. Hoeber, 1949), 423.

53. Tomlinson, "Interview with Robert Galbraith Heath." Heath was wrong about the impossibility of doing electrical stimulation work on the East Coast, as witnessed by the success of Yale neuroscientist José Delgado, whose experiments on patients from a Rhode Island psychiatric hospital began at the same time as the Tulane research.

54. Christina Fradelos, "The Last Desperate Cure: Electrical Brain Stimulation and Its Controversial Beginnings" (PhD diss., University of Chicago, 2008), 97, ProQuest (3309034).

55. Robert G. Heath, ed., *Studies in Schizophrenia: A Multidisciplinary Approach to Mind-Brain Relationships* (Cambridge, MA: Harvard University Press, 1954), 32.

56. Ibid., 350–51.

57. Ibid., 502, 535.

58. Ibid., 540.

59. Heinz Lehman and Thomas Ban, "History of the Psychopharmacology of Schizophrenia," *Canadian Journal of Psychiatry* 42 (1997): 152–62.

60. Adorno, *Authoritarian Personality*, v.

61. Stanley Milgram, *Obedience to Authority: An Experimental View* (New York: Harper and Row, 1974).

62. Thomas Blass, *The Man Who Shocked the World: The Life and Legacy of Stanley Milgram* (New York: Basic Books, 2004), 62.

63. Ibid. 268–69.

64. Ian Nicholson, "'Shocking' Masculinity: Stanley Milgram, 'Obedience to Authority' and the 'Crisis of Manhood' in Cold War America," *Isis* 102, no. 2 (2011). This is the most historically contextualized treatment of Milgram's experiment, but it does not explore the critical implications of the behaviorist setup.

65. Blass, *Man Who Shocked the World*, 77.

66. Skinner, *Matter of Consequences*, 242.

67. Robert G. Heath, ed., *The Role of Pleasure in Behavior* (New York: Harper and Row, 1964), 55.

68. Ibid., 57.

69. Ibid., 64.

70. Ibid., 62.

71. Ibid., 63.

72. Ibid., 70.

73. Ibid., 73.

74. Ibid., 78–79.

75. Cohen-Cole, *Open Mind*, 124.

76. Arthur Koestler, *The Ghost in the Machine* (New York: Macmillan, 1968), 262.

77. Arthur Koestler, *Darkness at Noon*, trans. Daphne Hardy (New York: Bantam Books, 1968), 208–10.

78. Koestler, *Ghost in the Machine*, 17.

79. Ibid., 353.

80. Immanuel Kant, *Groundwork of the Metaphysics of Morals*, trans. Mary Gregor (Cambridge: Cambridge University Press, 1998), 8.

81. Ibid., 37.

82. See Onora O'Neill, *Autonomy and Trust in Bioethics* (Cambridge: Cambridge University Press, 2002), 73–95, for a cogent and persuasive analysis of informed consent as a system of obligations rather than rights.

83. See Jerome Schneewind, "Autonomy after Kant," in Oliver Sensen, *Kant on Moral Autonomy* (Cambridge: Cambridge University Press, 2013), 146–68.

84. The notable exception was a behaviorist community in Mexico, under the strong leadership of a charismatic family patriarch. See Kuhlmann, *Living Walden Two*.

85. Skinner, *Matter of Consequences*, 317–21.

86. B. F. Skinner, *Beyond Freedom and Dignity* (London: Penguin, 1988), 36. John Stuart Mill's "On Liberty" was included in Skinner's despised literature of freedom. The supreme irony of including Mill in the roster of autonomy's naïve and superstitious defenders was that he might have been the first person to actually be raised according to behaviorist principles.

87. Ibid., 20.

88. Ibid., 76–77.

89. Ibid., 83.

90. Ibid., 126–42.

91. Robert G. Heath and Charles E. Moan, "Septal Stimulation for the Initiation of Heterosexual Behavior in a Homosexual Male," *Journal of Behavioral Therapy and Experimental Psychiatry* 3, no. 1 (1972): 23.

92. Rado, *Adaptational Psychodynamics*, 80, 78, 210–11.

93. Heath and Moan, "Septal Stimulation," 24–27.

94. Ibid., 27–28.

95. Ibid., 29.

96. Robert L. Spitzer, "The Diagnostic Status of Homosexuality in DSM-III: A Reformulation of the Issues," *American Journal of Psychiatry* 138, no. 2 (1981): 210–15.

97. Ibid., 211.

98. Noam Chomsky, "The Case against B. F. Skinner," *New York Review of Books*, December 30, 1971.

99. R. W. Robins, S. D. Gosling, and K. H. Craik, "An Empirical Analysis of Trends in Psychology," *American Psychology* 54, no. 2 (1999): 117–28.

100. Skinner, *Matter of Consequences*, 351–52.

101. Thomas Szasz, "Against Behaviorism: A Review of B. F. Skinner's *About Behaviorism*," *Libertarian Review* 111 (December 1974).

102. D. L. Rosenhan, "On Being Sane in Insane Places," *Science* 179, no 4070 (January 19, 1973): 250–58.

103. This fascinating episode, although tangentially related and much written about, lies beyond the scope of the present study.

104. Bill Rushton, "The Mysterious Experiments of Dr. Heath, In Which We Wonder Who Is Crazy and Who Is Sane," *Courier*, August 29–September 4, 1974, 5.

105. Ibid., 5.

106. Ibid., 7.

107. George J. Annas, *The Rights of Patients: The Basic ACLU Guide to Patient Rights* (1975; Carbondale: Southern Illinois University Press, 1989), 2.

Chapter 6

1. Duane Alexander and Patricia El-Hinnawy, "Interview with Duane Alexander: Belmont Oral History Project" (Office of Human Research Protections, US Department of Health and Human Services, 2004), 2.

2. Thomas Beauchamp, "The Belmont Report," April 18, 1979, http://www.hhs.gov/ohrp/humansubjects/guidance/belmont.html.

3. James Childress interviewed by Renee Fox and Judith Swazey, Project on Bioethics in American Society, "Interview with James Childress," Georgetown University line 456. https://repository.library.georgetown.edu/handle/10822/557025. See also Renée C. Fox and Judith P. Swazey, *Observing Bioethics* (New York: Oxford University Press, 2008), 21–76.

4. Thomas Beauchamp and James F. Childress, *Principles of Biomedical Ethics* (New York: Oxford University Press, 1979), 59.

5. Ibid., 65.

6. Jordan Goodman, Anthony McElligott, and Lara Marks, *Useful Bodies: Hu-*

mans in the Service of Medical Science in the Twentieth Century (Baltimore: Johns Hopkins University Press, 2003), chap. 8.

7. Beauchamp and Childress, *Principles of Biomedical Ethics*, 138.

8. Ibid., 11.

9. Ibid., 62–63.

10. Christopher Disman, "The San Francisco Bathhouse Battles of 1984," *Journal of Homosexuality* 44, nos. 3–4 (2003): 71–129.

11. Beth Ann Krier, "An Evening with Louise Hay: Controversial AIDS Counselor Draws Hundreds to Her Weekly Meetings and Message of Hope," *Los Angeles Times*, March 2, 1988.

12. Steven Epstein, *Impure Science: AIDS, Activism, and the Politics of Knowledge* (Berkeley: University of California Press, 1996), 190.

13. Ibid., 221.

14. Ibid., 195.

15. Peter S. Arno and Karen L. Feiden, *Against the Odds: The Story of AIDS Drug Development, Politics and Profits* (New York: Harper Collins, 1992), 43.

16. ACT UP, "Flyer of the First ACT UP Action, March 24, 1987, Wall Street, New York City." http://www.actupny.org/documents/1stFlyer.html.

17. Richard L. Landau, ed., *Regulating New Drugs* (Chicago: University of Chicago, Center for Policy Study, 1973).

18. Edward Nik-Khah, "Neoliberal Pharmaceutical Science and the Chicago School of Economics," *Social Studies of Science* 44, no. 4 (August 1, 2014): 489–517.

19. "New Initiatives to Speed Access to New Drugs," *AIDS Info*, April 9, 1992. https://aidsinfo.nih.gov/news/223/new-initiatives-to-speed-access-to-new-drugs.

20. Lucas Lichert, *Conservatism, Consumer Choice and the Food and Drug Administration during the Reagan Era: a Prescription for Scandal* (Lanham, MD: Lexington Books, 2014), 145.

21. "Approval of Zalcitabine," *AIDS Info*, June 22, 1992. https://aidsinfo.nih.gov/news/29/approval-of-zalcitabine.

22. For example, the blockbuster cholesterol controlling drugs statins were approved on the basis of cholesterol tests rather than longevity or heart-attack data, and have made billions of dollars for their makers without necessarily producing any improvement in patients' health. J. Lenzer, "Unreported Cholesterol Drug Data Released by Company," *British Medical Journal* 336, no. 7637 (2008): 180–81.

23. Martin Delaney interviewed for a PBS *Frontline* special "The Age of AIDS," 2006. http://www.pbs.org/wgbh/pages/frontline/aids/interviews/delaney.html.

24. Michie Hunt, "Direct to Consumer Advertising of Prescription Drugs," *National Health Policy Forum* (1998), http://www.nhpf.org/library/background-papers/bp_dtc_4-98.pdf.

25. William Stanley Jevons, *The Theory of Political Economy*, 2nd ed. (London: Macmillan, 1879), vii. With thanks to Chad Valasek, for setting me on the track of utilitarian behavioral economics then and now.

26. Jevons was inspired by an earlier work of utilitarian psychology, Richard Jennings, *Natural Elements of Political Economy* (London: Longman, Brown, Green

and Longmans, 1855), which defined the discipline of political economy as dealing with only those human traits shared with the "brute creation" and "only those motives which are derived, more or less remotely, from the attraction of pleasure or the repulsion of pain" (45).

27. Jevons, *Theory of Political Economy*, 11.

28. Ibid., 35.

29. Ibid., 12.

30. Ibid., 39.

31. Francis Ysidro Edgeworth, *Mathematical Psychics: An Essay on the Application of Mathematics to the Moral Sciences* (London: C. Kegan Paul, 1881), 98, 101–2.

32. Ibid., 97–98.

33. Ibid., 104.

34. Perhaps the most influential discussion of this problem is to be found in the work of the economist, sociologist, and philosopher Vilfredo Pareto, for example, *The Mind and Society*, vol 4., *The General Form of Society*, trans. Andrew Bongiorno and Arthur Livingston (New York: Harcourt, Brace and Co., 1935), 1456–83.

35. For the whole story from Bentham to the present, see Daniel Read, "Experienced Utility: Utility Theory from Jeremy Bentham to Daniel Kahneman," *Thinking & Reasoning* 13, no. 1 (2007): 45–61.

36. Irving Fisher, "Is 'Utility' the Most Suitable Term for the Concept It Is Used to Denote?" *American Economic Review* 8, no. 2 (June 1918): 335. My thanks to Theo Dryer for sending me this reference.

37. Friedrich A. Hayek et al., *The Road to Serfdom: The Condensed Version of the Road to Serfdom by F. A. Hayek as It Appeared in the April 1945 Edition of Reader's Digest* (London: Institute of Economic Affairs, 2001), 53. Examples of the same anti-utilitarian arguments in the period multiply; see, for example, Karl R. Popper, *The Open Society and Its Enemies* (Princeton, NJ: Princeton University Press, 1966), especially the chapter on "Philosopher Kings," or the chapter on Helvétius in Isaiah Berlin, *Freedom and Its Betrayal: Six Enemies of Human Liberty* (London: Chatto and Windus, 2002).

38. As one historian of the relation between psychology and consumer choice theory has recently remarked, "Theories that provide a mechanical or non-volitional characterisation of consumer behaviour—either psychophysiology or strict behaviourism—do not support this notion of free choice and thus simply will not do." D. Wade Hands, "Economics, Psychology and the History of Consumer Choice Theory," *Cambridge Journal of Economics* 34, no. 4 (July 1, 2010): 633–48.

39. Daniel Kahneman and Amos Tversky, "Prospect Theory: An Analysis of Decision under Risk," *Econometrica* 42 (1979): 263.

40. Richard H. Thaler, *Misbehaving: The Making of Behavioral Economics* (New York: W. W. Norton, 2015), footnote on 34.

41. Daniel Kahneman, Ed Diener, and Norbert Schwarz, eds., *Well-being: The Foundations of Hedonic Psychology* (New York: Russell Sage Foundation, 1999), 22.

42. For a trenchant critique of this approach, see Roberto Fumagalli, "The Futile Search for True Utility," *Economics and Philosophy* 29, no. 3 (2013): 325–47.

43. Richard J. Herrnstein, "Behavior, Reinforcement and Utility," in *The Origin of Values*, ed. Michael Hechter, Lynn Nadel, and Richard E. Michod (New York: A. de Gruyter, 1993), 137.

44. Ibid., 150.

45. Paul W. Glimcher, Michael C. Dorris, and Hannah M. Bayer, "Physiological Utility Theory and the Neuroeconomics of Choice," *Games and Economic Behavior* 52, no. 2 (2005): 218, 254.

46. B. F. Skinner, *The Shaping of a Behaviorist* (New York: Alfred A. Knopf, 1979), 197.

47. Gary S. Becker and Kevin M. Murphy, "A Theory of Rational Addiction," *Journal of Political Economy* 96, no. 4 (August 1988): 675.

48. Ibid., 681–82.

49. Hands, *Economics, Psychology*, 633–48.

50. Walter Mischel, "Preference for Delayed Reinforcement: An Experimental Study of a Cultural Observation," *Journal of Abnormal and Social Psychology* 56, no. 1 (January 1958): 57.

51. Walter Mischel and Ebbe B. Ebbesen, "Attention in Delay of Gratification," *Journal of Personality and Social Psychology* 16, no. 2 (1970): 335.

52. Yiuchi Shoda, Walter Mischel, and Philip K. Peake, "Predicting Adolescent Cognitive and Self-Regulatory Competencies from Preschool Delay of Gratification: Identifying Diagnostic Conditions," *Developmental Psychology* 26, no. 6 (1990): 985.

53. Richard H. Thaler and H. M. Shefrin, "An Economic Theory of Self-Control," *Journal of Political Economy* 89, no. 2 (April 1981): 394.

54. Mark Egan, "Nudge Database," Stirling Behavioral Science Centre, https://www.stir.ac.uk/media/schools/management/documents/economics/Nudge%20Database%201.2.pdf.

55. Daniel Kahneman, *Thinking, Fast and Slow* (New York: Farrar, Strauss and Giroux, 2011), 377.

56. Daniel Kahneman, Peter P. Wakker, and Rakesh Sarin, "Back to Bentham? Explorations of Experienced Utility," *Quarterly Journal of Economics* 112, no. 2 (May 1997): 375.

57. Joshua Greene, *Moral Tribes: Emotion, Reason, and the Gap between Us and Them* (New York: Penguin, 2013), 189.

58. Ibid., 186.

59. Ibid., 198–99.

60. Ibid., 308.

61. Ibid., 128–31; ellipses in the original.

62. Hugo Adam Bedau, "Anarchical Fallacies: Bentham's Attack on Human Rights," *Human Rights Quarterly* 22, no. 1 (2000): 278.

63. Thaler, *Misbehaving*, 87.

64. Adam Smith, *The Theory of Moral Sentiments*, ed. D. D. Raphael and A. L. Macfie (Oxford: Clarendon Press, 1979), 189.

65. This exhortation applies equally to Smith's economic writings. I first read *The Wealth of Nations* during the period when Occupy Wall Street was making headlines,

and I was struck, like so many before me, by what a die-hard champion of the 99 percent Smith was, always railing against the way that capitalists conspire to raise prices, lower wages, and prevent workers from organizing. His focus was at least as much on social justice as it was on liberty and prosperity, and I can only suppose that the free marketeers who invoke him, as well as the leftists who denounce him, have never read him. For a learned reclamation of Adam Smith for the left of center, see Emma Rothschild, *Economic Sentiments: Adam Smith, Condorcet and the Enlightenment* (Cambridge, MA: Harvard University Press, 2001).

Afterword

1. My thanks to psychotherapist and fellow sufferer Hallie Kushner for this insightful diagnosis.

2. Charles Darwin, *The Descent of Man and Selection in Relation to Sex* (New York: D. Appleton and Company, 1871), 162.

3. Thomas Beddoes, *Contributions to Physical and Medical Knowledge, Principally from the West of England* (London: T. M. Longman and O. Rees, 1799), 4.

4. Robert Zubrin "How Scientifically Accurate Is The Martian?" *Guardian*, October 6, 2015, https://www.theguardian.com/film/2015/oct/06/how-scientifically -accurate-is-the-martian.

5. Isaac T. Yonometo et al., "Dual Organism Design Cycle Reveals Small Subunit Substitutions That Improve [NiFe] Hydrogenase Hydrogen Evolution," *Journal of Biological Engineering* 7, no. 17 (2013).

6. Robert Friedman, J. Craig Venter Institute, La Jolla, personal communication, June 6, 2016.

7. Herbert Spencer, "A Theory of Population, Deduced from the General Law of Animal Fertility," *Westminster Review* 57, no. 112 (1852): 501.

8. My thanks to William Peden for this pithy formulation.

9. Richard Hofstadter, *Social Darwinism in American Thought* (New York: George Brazillier, 1959), 35.

10. David Duncan, *The Life and Letters of Herbert Spencer* (London: Methuen, 1908), 410.

11. For the invocations of human rights, see Samuel Moyn, *The Last Utopia: Human Rights in History* (Cambridge, MA: Harvard University Press, 2012).

12. Herbert Spencer, *The Data of Ethics* (London: G. Norman and Son, 1879), 46.

13. Ibid., 68–69.

14. Ibid., 261–64.

BIBLIOGRAPHY

Adorno, Theodor W. *The Authoritarian Personality*. New York: Harper, 1950.

Alexander, Duane, and Patricia El-Hinnawy. "Interview with Duane Alexander: Belmont Oral History Project." Office of Human Research Protections, US Department of Health and Human Services, 2004.

Alexander, Shana. "They Decide Who Lives, Who Dies." *Life* 53, no. 19 (November 9, 1962): 103.

Annas, George J. *The Rights of Patients: The Basic ACLU Guide to Patient Rights*. 1975; Carbondale: Southern Illinois University Press, 1989.

Annas, George J., and Michael A. Grodin. *The Nazi Doctors and the Nuremberg Code: Human Rights in Human Experimentation*. New York: Oxford University Press, 1992.

Armitage, David. *The Declaration of Independence: A Global History*. Cambridge, MA: Harvard University Press, 2007.

Arno, Peter S., and Karen L. Feiden. *Against the Odds: The Story of AIDS Drug Development, Politics and Profits*. New York: Harper Collins, 1992.

Atkinson, Charles Milner. *Jeremy Bentham: His Life and Work*. London: Methuen & Co., 1905.

Aubrey, John. *Brief Lives, Chiefly of Contemporaries, Set Down by John Aubrey, between the Years 1669 and 1696*. Edited by Andrew Clark. Oxford: Clarendon Press, 1898.

Bain, Alexander. *Autobiography*. London: Longmans, Green & Co., 1904.

———. *The Emotions and the Will*. London: Longmans, Green & Co., 1875.

———. *James Mill: A Biography*. London: Longmans, Green & Co., 1882.

——. *John Stuart Mill: A Criticism with Personal Reflections*. London: Longmans, Green & Co., 1882.

——. *The Senses and the Intellect*. London: Longmans, Green & Co., 1868.

Baldwin, James Mark. *Mental Development in the Child and the Race: Methods and Processes*. New York: Macmillan and Co., 1895.

Ballance, Charles A. *A Glimpse into the History of Surgery of the Brain*. London: Macmillan and Co., 1922.

Bartholow, Roberts. "Experimental Investigations into the Functions of the Human Brain." *American Journal of the Medical Sciences* 84 (April 1874): 305–13.

Beauchamp, Tom, and James F. Childress. *Principles of Biomedical Ethics*. New York: Oxford University Press, 1979.

Beauchamp, Tom, and Patricia El-Hinnawy. "Interview with Tom Lamar Beauchamp: Belmont Oral History Project." Office of Human Research Protections, US Department of Health and Human Services, 2004.

Beccaria, Cesare. *An Essay on Crimes and Punishments*. London: J. Almon, 1767.

Becker, Gary S., and Kevin M. Murphy. "A Theory of Rational Addiction." *Journal of Political Economy* 96, no. 4 (August 1988): 675–700.

Bedau, Hugo Adam. "Anarchical Fallacies: Bentham's Attack on Human Rights." *Human Rights Quarterly* 22, no. 1 (2000): 261–79.

Beddoes, Thomas. *Contributions to Physical and Medical Knowledge, Principally from the West of England*. London: T. M. Longman and O. Rees, 1799).

Beecher, Henry. "Ethics and Clinical Research." In *Ethics in Medicine: Historical Perspectives and Contemporary Concerns*, edited by Stanley Joel Reiser, Arthur J. Dyck and William J. Curran, 288–93. Cambridge, MA: MIT Press, 1977.

——. "Experimentation in Man." *Journal of the American Medical Association* 169, no. 5 (1959): 461–78.

Bell, Charles. *Letters of Sir Charles Bell, Selected from his Correspondence with his Brother George Joseph Bell*. London: John Murray, 1870.

Bentham, Jeremy. *Bentham's Auto-Icon and Related Writings*. Edited by James E. Crimmins. Bristol: Thoemmes Press, 2002.

——. *The Correspondence of Jeremy Bentham*. Vol. 3, *January 1781 to October 1788*. Edited by Ian Christie. London: Athlone Press, 1971.

——. *The Correspondence of Jeremy Bentham*. Vol. 4, *October 1788 to December 1793*. Edited by Alexander Taylor Milne. London: Athlone Press, 1981.

——. *An Introduction to the Principles of Morals and Legislation*. Edited by J. H. Burns and H. L. A. Hart. Oxford: Clarendon Press, 1996.

——. *The Panopticon Writings*. Edited by Miran Božovič. London: Verso, 1995.

——. *Plan of Parliamentary Reform, in the form of a Catechism*. London: R. Hunter, 1817.

——. *Rights, Representation and Reform: Nonsense upon Stilts and Other Writings on the French Revolution*. Edited by Philip Scholfield, Catherine Pease-Watkin, and Cyprian Blamires. Oxford: Clarendon Press, 2002.

——. *The Works of Jeremy Bentham*. Edited by John Bowring New York: Russell & Russell, 1962.

Benya, Frazier. "Biomedical Advances Confront Society: Congressional Hearings and the Development of Bioethics, 1960–1975." PhD, University of Minnesota, 2012.

Berlin, Isaiah. *Freedom and Its Betrayal: Six Enemies of Human Liberty*. London: Chatto and Windus, 2002.

Blass, Thomas. *The Man Who Shocked the World: The Life and Legacy of Stanley Milgram*. New York: Basic Books, 2004.

Boakes, Robert. *From Darwin to Behaviorism: Psychology and the Minds of Animals*. Cambridge: Cambridge University Press, 1984.

Bonin, Gerhardt von. *Some Papers on the Cerebral Cortex, Translated from the French and German*. Springfield, IL: Thomas, 1960.

Brandt, Allan M. "Racism and Research: The Case of the Tuskegee Syphilis Study." *Hastings Center Report* 8, no. 6 (December 1978): 21–29.

Bretschneider, Hubert. *Der Streit um die Vivisektion im 19. Jahrhundert*. Stuttgart: Gustav Fischer, 1962.

Brody, Jane E. "All in the Name of Science." *New York Times*, July 30, 1972.

Brower, Daniel. "The Problem of Quantification in Psychological Science." *Psychological Review* 56, no. 6 (1949): 325–33.

Buckley, Kerry W. *Mechanical Man: John Broadus Watson and the Beginnings of Behaviorism*. New York: Guilford Press, 1989.

Bush, Vannevar. *Science, the Endless Frontier: A Report to the President on a Program for Postwar Scientific Research*. Washington, DC: Government Printing Office, 1945.

Capaldi, Nicholas. *John Stuart Mill: A Biography*. Cambridge: Cambridge University Press, 2004.

Capshew, James. "Engineering Behavior: Project Pigeon, World War II, and the Conditioning of B. F. Skinner." In *B. F. Skinner and Behaviorism in America Culture*, edited by Laurence Smith and William Woodward, 128–50. London: Associated University Presses, 1996.

"The Charge against Professor Ferrier under the Vivisection Act: Dismissal of the Summons." *British Medical Journal* 2, no. 1090 (November 19, 1881): 836–42.

Chesney, A. M., and J. E. Kemp. "Studies in Experimental Syphilis: VI. On Variations in the Response of Treated Rabbits to Reinoculation and on Cryptogenetic Reinfection with Syphilis." *Journal of Experimental Medicine* 44, no. 5 (1926): 589–606.

Childress, James. "Who Shall Live When Not All Can Live." In *Ethics in Medicine, Historical Perspectives and Contemporary Concerns*, edited by Stanley Joel Reiser, Arthur J. Dyck, and William J. Curran, 620–26. Cambridge, MA: MIT Press, 1977.

Chomsky, Noam. "The Case against B. F. Skinner." *New York Review of Books*, December 30, 1971.

Crichton-Browne, James. "Medical Director's Journal: Reports to the Quarterly Meetings, 1867–1874."

———. *Report of the Committee of Visitors and of the Medical Superintendent of the*

West Riding Pauper Lunatic Asylum for the Year 1868. Wakefield: Hicks and Allen, 1869.

———. *Report of the Committee of Visitors and of the Medical Superintendent of the West Riding Pauper Lunatic Asylum for the Year 1871*. Wakefield: B. W. Allen, 1872.

Cohen-Cole, Jamie. *The Open Mind: Cold War Politics and the Sciences of Human Nature*. Chicago: University of Chicago Press, 2014.

Compston, Alastair. "Editorial." *Brain* 134 (2011): 1251–53.

Cooke, Robert, and Patricia El-Hinnawy. "Interview with Robert E. Cooke: Belmont Oral History Project." Office of Human Research Protections, US Department of Health and Human Services, 2004.

Cowherd, Raymond G. *Political Economists and the English Poor Laws: A Historical Study of the Influence of Classical Economics on the Formation of Social Welfare Policy*. Athens: Ohio University Press, 1977.

Critchley, Macdonald, and Eileen A. Critchley. *John Hughlings Jackson: Father of English Neurology*. New York: Oxford University Press, 1998.

Cunniffe, E. R. "Supplementary Report of the Judicial Council." *Journal of the American Medical Association* 132, no. 17 (December 28, 1946): 1090.

Cutler, John C. "Final Syphilis Report." Unpublished research report. Department of Health, Education and Welfare, Public Health Service, Venereal Disease Division, 1948. Record Group 442: Records of the Centers for Disease Control and Prevention, 1921–2006, National Archives: 6. https://nara-media-001.s3 .amazonaws.com/arcmedia/research/health/cdc-cutler-records/folder-1-syphilis -report.pdf.

Darwin, Charles. *The Descent of Man and Selection in Relation to Sex*. New York: D. Appleton and Company, 1871.

———. *The Expression of the Emotions in Man and Animals*. New York: D. Appleton and Company, 1886.

———. *The Life and Letters of Charles Darwin, Including an Autobiographical Chapter*. Edited by Francis Darwin. Vol. 1. London: John Murray, 1888.

Daunton, M. J. *Progress and Poverty: An Economic and Social History of Britain, 1700–1850*. Oxford: Oxford University Press, 1995.

Davy, Humphry. *Researches, Chemical and Philosophical; Chiefly Concerning Nitrous Oxide or Dephlogisticated Nitrous Air, and Its Respiration*. London: J. Johnson, 1800.

Disman, Christopher. "The San Francisco Bathhouse Battles of 1984." *Journal of Homosexuality* 44, nos. 3–4 (2003): 71–129.

"The Drama in Lunacy Treatment." *Medical Times and Gazette*, April 17, 1875.

"Dr. Ferrier's Localizations: For Whose Advantage?" *British Medical Journal* 2, no. 1090 (November 19, 1881): 822–24.

Duncan, David. *The Life and Letters of Herbert Spencer*. London: Methuen, 1908.

Edgeworth, Francis Ysidro. *Mathematical Psychics: An Essay on the Application of Mathematics to the Moral Sciences* (London: C. Kegan Paul, 1881), 98, 101–2.

Ellis, William Charles. *A Treatise on the Nature, Symptoms, Causes, and Treatment of Insanity*. London: Holdsworth, 1838.

Epstein, Steven. *Impure Science: AIDS, Activism, and the Politics of Knowledge*. Berkeley: University of California Press, 1996.

"Ethically Impossible: STD Research in Guatemala from 1946 to 1948." Washington, DC: Presidential Commission for the Study of Bioethical Issues, 2011. http://bioethics.gov/node/654

Evans, John Hyde. *The History and Future of Bioethics: A Sociological View*. New York: Oxford University Press, 2012.

Evans, Robin. *The Fabrication of Virtue: English Prison Architecture, 1750–1840*. Cambridge: Cambridge University Press, 1982.

Ferrier, David. "Bartholow's Experiments on the Human Brain." *London Medical Record* 2 (1874): 285–86.

———. "Experiments on the Brains of Monkeys: First Series." *Proceedings of the Royal Society of London* 23 (1874–1875): 409–30.

———. "Experiments on the Brains of Monkeys: Second Series." *Philosophical Transactions of the Royal Society of London* 165 (1875): 433–88.

———. *The Functions of the Brain*. London: Smith, Elder & Co., 1876.

Ferster, C. B., and B. F. Skinner. *Schedules of Reinforcement*. East Norwalk, CT: Appleton-Century-Crofts, 1957.

"The Festivities of the Congress." *British Medical Journal* 2, no. 1076 (August 13, 1881): 303–4.

Final Report of the Advisory Committee on Human Radiation Experiments. New York: Oxford University Press, 1996.

Finger, Stanley. *Minds behind the Brain*. Oxford: Oxford University Press, 2000.

Fisher, Irving. "Is 'Utility' the Most Suitable Term for the Concept It Is Used to Denote?" *American Economic Review* 8, no. 2 (June 1918): 335–37.

Fradelos, Christina. "The Last Desperate Cure: Electrical Brain Stimulation and Its Controversial Beginnings." PhD diss., University of Chicago, 2008.

Fumagalli, Roberto. "The Futile Search for True Utility." *Economics and Philosophy* 29, no. 3 (2013): 325–47.

Glimcher, Paul W., Michael C. Dorris, and Hannah M. Bayer. "Physiological Utility Theory and the Neuroeconomics of Choice." *Games and Economic Behavior* 52, no. 2 (2005): 213–56.

Goodman, Jordan, Anthony McElligott, and Lara Marks. *Useful Bodies : Humans in the Service of Medical Science in the Twentieth Century*. Baltimore: Johns Hopkins University Press, 2003.

Gray, Fred. *The Tuskegee Syphilis Study: The Real Story and Beyond*. Montgomery, AL: NewSouth Books, 1998.

Green, Michele. "Sympathy and Self-Interest: The Crisis in Mill's Mental History." *Utilitas* 1 (1989): 259–77.

Greene, Joshua David. *Moral Tribes: Emotion, Reason, and the Gap between Us and Them*. New York: Penguin, 2013.

Halbach, Erich Hans, and Hans O. Luxenburger. "Experiments on Human Beings as Viewed in World Literature." Becker-Freyseng Document Book IV, document no. 60 (1947), in Harvard Law School Library, "Nuremberg Trials Project," http://nuremberg.law.harvard.edu/documents/229. (Also, "Supplement to the Report," document no. 60a, http://nuremberg.law.harvard.edu/documents/230.)

Haliburton, Rachel. *Autonomy and the Situated Self: A Challenge to Bioethics*. Lanham, MD: Lexington Books, 2013.

Halpern, Sydney A. *Lesser Harms: The Morality of Risk in Medical Research*. Chicago: University of Chicago Press, 2004.

Hands, D. Wade. "Economics, Psychology and the History of Consumer Choice Theory." *Cambridge Journal of Economics* 34, no. 4 (July 1, 2010): 633–48.

Hare, R. M. *Moral Thinking: Its Levels, Method, and Point*. Oxford: Oxford University Press, 1981.

Harkness, Jon M. "Nuremberg and the Issue of Wartime Experiments on US Prisoners." *Journal of the American Medical Association* 276, no. 20 (1996): 1672–75.

Harris, Lauren Julius, and Jason B. Almerigi. "Probing the Human Brain with Stimulating Electrodes: The Story of Roberts Bartholow's (1874) Experiment on Mary Rafferty." *Brain and Cognition* 70, no. 1 (2009): 92–115.

Harris, Richard. *The Real Voice*. New York: Macmillan, 1964.

Hartley, David. *Observations on Man: His Frame, His Duties and His Expectations*. 2nd ed. London: J. Johnson, 1791.

Hayek, Friedrich A. von, Edwin J. Feulner, John Blundell, and the Institute of Economic Affairs (Great Britain). *The Road to Serfdom: The Condensed Version of the Road to Serfdom by F. A. Hayek as It Appeared in the April 1945 Edition of Reader's Digest*. London: Institute of Economic Affairs, 2001.

Heath, Robert G. *Exploring the Mind-Brain Relationship*. Baton Rouge, LA: Moran Printing, 1996.

———, ed. *The Role of Pleasure in Behavior; a Symposium by 22 Authors*. New York: Hoeber-Harper, 1964.

———, ed. *Studies in Schizophrenia: A Multidisciplinary Approach to Mind-Brain Relationships*. Cambridge, MA: Harvard University Press, 1954.

Heath, Robert G., and Charles E. Moan. "Septal Stimulation for the Initiation of Heterosexual Behavior in a Homosexual Male." *Journal of Behavioral Therapy and Experimental Psychiatry* 3, no. 1 (1972): 23–30.

Heller, Jean. "Syphilis Victims in U.S. Study Went Untreated for 40 Years." *New York Times*, July 26 1972.

Helvétius, Claude. *De l'Esprit; or, Essays on the Mind, and its several Faculties*. Translated by William Mudford. London: M. Jones, 1807.

Herrnstein, Richard J. "Behavior, Reinforcement and Utility." In *The Origin of Values*, edited by Michael Hechter, Lynn Nadel, and Richard E. Michod. New York: A. de Gruyter, 1993.

Himmelfarb, Gertrude. "Bentham's Utopia: The National Charity Company." *Journal of British Studies* 10, no. 1 (November 1970): 80–125.

Hobbes, Thomas. *The Correspondence of Thomas Hobbes*. Edited by Noel Malcolm. Oxford: Oxford University Press, 1994.

———. *De Cive; or, The Citizen*. Edited by Sterling Lamprecht. New York: Appleton-Century-Crofts, 1949.

———. *Leviathan; or, The Matter, Form and Power of the Commonwealth, Ecclesiastical and Civil*. Edited by C. B. Macpherson. Harmondsworth: Penguin, 1968.

Hofstadter, Richard. *Social Darwinism in American Thought*. New York: George Brazillier, 1959.

Hooper, Judith, and Teresi, Dick. *The Three-Pound Universe*. New York: Macmillan, 1986.

Howell, Thomas Bayly. *Cobbett's Complete Collection of State Trials, and Proceedings for High Treason and Other Crimes and Misdemeanours, from the Earliest Period to the Present Time*. London: Hansard, 1809.

Hume, David. *Philosophical Works*. Bristol: Thoemmes Press, 1996.

Hunt, Michie. "Direct to Consumer Advertising of Prescription Drugs." *National Health Policy Forum* (1998). http://www.nhpf.org/library/background-papers/bp_dtc_4-98.pdf

Hutcheson, Francis. *Inquiry into the Original of our Ideas of Beauty and Virtue, in Two Treatises*. London: Midwinter, 1738.

Jackson, John Hughlings. *Selected Writings of John Hughlings Jackson: On Epilepsy and Epileptiform Convulsions*. London: Hodder and Stoughton, 1931.

Jay, Mike. *The Atmosphere of Heaven: The Unnatural Experiments of Dr. Beddoes and His Sons of Genius*. New Haven, CT: Yale University Press, 2009.

Jennings, Richard. *Natural Elements of Political Economy*. London: Longman, Brown, Green and Longmans, 1855.

Jevons, William Stanley. *The Theory of Political Economy*. 2nd ed. London: Macmillan, 1879.

Jonçich, Geraldine M. *The Sane Positivist a Biography of Edward L. Thorndike*. Middletown, CT: Wesleyan University Press, 1968.

Jones, Howard. *The Epicurean Tradition*. London: Routledge, 1989.

Kahneman, Daniel. *Thinking, Fast and Slow*. New York: Farrar, Strauss and Giroux, 2011.

Kahneman, Daniel, Ed Diener, and Norbert Schwarz, eds. *Well-being: The Foundations of Hedonic Psychology*. New York: Russell Sage Foundation, 1999.

Kahneman, Daniel, and Amos Tversky. "Prospect Theory: An Analysis of Decision under Risk." *Econometrica* 42 (1979): 263–91.

Kahneman, Daniel, Peter P. Wakker, and Rakesh Sarin. "Back to Bentham? Explorations of Experienced Utility." *Quarterly Journal of Economics* 112, no. 2 (May 1997): 375–405.

Kant, Immanuel. *Groundwork of the Metaphysics of Morals*. Translated by Mary Gregor. Cambridge: Cambridge University Press, 1998.

Katz, Jay, Alexander Morgan Capron, and Eleanor Swift Glass. *Experimentation with Human Beings; the Authority of the Investigator, Subject, Professions, and*

State in the Human Experimentation Process. New York: Russell Sage Foundation, 1972.

Killen, Andreas. *1973 Nervous Breakdown: Watergate, Warhol, and the Birth of Post-Sixties America*. New York: Bloomsbury, 2006.

Koestler, Arthur. *Darkness at Noon*. Translated by Daphne Hardy. New York: Bantam Books, 1968.

———. *The Ghost in the Machine*. New York: Macmillan, 1968.

Krier, Beth Ann. "An Evening with Louise Hay: Controversial AIDS Counselor Draws Hundreds to Her Weekly Meetings and Message of Hope." *Los Angeles Times*, March 2, 1988.

Kuhlmann, Hilke. *Living Walden Two: B. F. Skinner's Behaviorist Utopia and Experimental Communities*. Urbana: University of Illinois Press, 2005.

Laitinen, Lauri. "Ethical Aspects of Psychiatric Surgery." In *Neurosurgical Treatment in Psychiatry, Pain, and Epilepsy*, edited by William Herbert Sweet et al., 483–88. Baltimore: University Park Press, 1977.

Landau, Richard L., ed. "Regulating New Drugs." Chicago: University of Chicago, Center for Policy Study, 1973.

Langer, Elinor. "Human Experimentation: Cancer Studies at Sloan-Kettering Stir Public Debate on Medical Ethics." *Science* 143, no. 3606 (1964).

Latour, Bruno. *The Pasteurization of France*. Cambridge, MA: Harvard University Press, 1988.

Lederer, Susan E. *Subjected to Science: Human Experimentation in America before the Second World War*. Baltimore: Johns Hopkins University Press, 1995.

Lehman, Heinz, and Thomas Ban. "History of the Psychopharmacology of Schizophrenia." *Canadian Journal of Psychiatry* 42 (1997): 152–62.

Lenzer, J. "Unreported Cholesterol Drug Data Released by Company." *British Medical Journal* 336, no. 7637 (2008): 180–81.

Lichert, Lucas. *Conservatism, Consumer Choice and the Food and Drug Administration during the Reagan Era: A Prescription for Scandal*. Lanham, MD: Lexington Books, 2014.

Locke, John. *An Essay Concerning Human Understanding*. Edited by Peter Nidditch. Oxford: Oxford University Press, 1975.

———. *An Essay Concerning Toleration and Other Writings on Law and Politics, 1667–1683*. Edited by J. R. Milton and Philip Milton. Oxford: Clarendon Press, 2006.

———. *Two Tracts on Government*. Edited by Philip Abrams. Cambridge: Cambridge University Press, 1967.

———. *Two Treatises of Government*. Edited by Peter Laslett. 2nd ed. Cambridge: Cambridge University Press, 1970.

Malthus, T. R. *An Essay on the Principle of Population; or, A View of Its Past and Present Effects on Human Happiness; with an Inquiry into Our Prospects Respecting the Future Removal or Mitigation of the Evils Which It Occasions*. London: J. Johnson, 1803.

————. *The Works of Thomas Robert Malthus.* Edited by E. A. Wrigley and David Souden. London: William Pickering, 1986.

Martinich, A. P. *Hobbes: A Biography.* Cambridge: Cambridge University Press, 1999.

Mazlish, Bruce. *James and John Stuart Mill: Father and Son in the Nineteenth Century.* 2nd ed. New Brunswick, NJ: Transaction Inc., 1988.

McCance, R. A. "The Practice of Experimental Medicine." *Proceedings of the Royal Society of Medicine* 44 (1951): 189–94.

Mettler, Fred, ed. *Selective Partial Ablation of the Frontal Cortex.* New York: Paul B. Hoeber, 1949.

Milgram, Stanley. *Obedience to Authority: An Experimental View.* New York: Harper and Row, 1974.

Mill, James. *Analysis of the Phenomena of the Human Mind.* London: Longmans, 1869.

————. *The History of British India.* London: James Madden & Co., 1840.

Mill, John Stuart. *Collected Works of John Stuart Mill.* Edited by John M. Robson and Jack Stillinger. London: Routledge, 1981.

Milton, J. R. "Locke's Life and Times." In *The Cambridge Companion to Locke*, edited by Vere Chappell, 5–25. Cambridge: Cambridge University Press, 1997.

Mintz, Samuel. *The Hunting of Leviathan: Seventeenth-Century Reactions to the Materialism and Moral Philosophy of Thomas Hobbes.* Cambridge: Cambridge University Press, 2010.

Mischel, Walter. "Preference for Delayed Reinforcement: An Experimental Study of a Cultural Observation." *Journal of Abnormal and Social Psychology* 56, no. 1 (1958): 57–61.

Mischel, Walter, and Ebbe B. Ebbesen. "Attention in Delay of Gratification." *Journal of Personality and Social Psychology* 16, no. 2 (1970): 329–37.

Moreno, Jonathan D., and Susan E. Lederer. "Revising the History of Cold War Research Ethics." *Kennedy Institute of Ethics Journal* 6, no. 3 (1996): 223–37.

Morgan, Conwy Lloyd. *Animal Life and Intelligence.* Boston: Ginn and Co., 1891.

————. *Habit and Instinct.* London: E. Arnold, 1896.

————. *An Introduction to Comparative Psychology.* London: Walter Scott, 1894.

Moyn, Samuel. *The Last Utopia: Human Rights in History.* Cambridge, MA: Harvard University Press, 2012.

Nicholls, George. *A History of the English Poor Law, in Connection with the State of the Country and the Condition of the People.* 2nd ed. Vol. 2. New York: G. P. Putnam and Sons, 1898.

Nicholson, Ian. "'Shocking' Masculinity: Stanley Milgram, 'Obedience to Authority' and the 'Crisis of Manhood' in Cold War America." *Isis* 102, no. 2 (2011): 238–68.

Nik-Khah, Edward. "Neoliberal Pharmaceutical Science and the Chicago School of Economics." *Social Studies of Science* 44, no. 4 (August 1, 2014): 489–517.

Olds, J., and P. Milner. "Positive Reinforcement Produced by Electrical Stimulation of the Septal Area and Other Regions of Rat Brain." *Journal of Comparative and Physiological Psychology* 47, no. 6 (1954): 419–27.

O'Neill, Onora. *Autonomy and Trust in Bioethics*. Cambridge: Cambridge University Press, 2002.

Oppenheim, Janet. *"Shattered Nerves": Doctors, Patients, and Depression in Victorian England*. New York: Oxford University Press, 1991.

Oshinsky, David M. *Polio: An American Story*. Oxford: Oxford University Press, 2005.

Osmundsen, John A. "Physician Scores Tests on Humans: He Asserts Experiments Are Done without Consent." *New York Times*, March 24, 1965.

Paley, William. *Principles of Moral and Political Philosophy*. Boston: West and Richardson, 1815.

Pareto, Vilfredo. *The Mind and Society*. Vol. 4, *The General Form of Society*. Translated by Andrew Bongiorno and Arthur Livingston. New York: Harcourt, Brace and Co., 1935.

Phillipson, Nicholas. *Adam Smith: An Enlightened Life*. London: Penguin, 2010.

Popper, Karl R. *The Open Society and Its Enemies*. Princeton, NJ: Princeton University Press, 1966.

Potter, Robert. *Observation on the Poor Laws, on the Present State of the Poor, and Houses of Industry*. London: J. Wilkie, 1775.

Priestley, Joseph. *Political Writings*. Edited by Peter Miller. Cambridge: Cambridge University Press, 1993.

Proctor, Robert. *Racial Hygiene: Medicine under the Nazis*. Cambridge, MA: Harvard University Press, 1988.

Quality of Health Care—Human Experimentation: Hearings before the Committee on Labor and Public Welfare, Subcommittee on Health. 93rd Cong. (1973).

Rado, Sandor. *Adaptational Psychodynamics*. Edited by Jean Jameson and Henriette Klein. Northvale, NJ: Jason Aronson, 1995.

Rae, John. *Life of Adam Smith*. London: Macmillan, 1895.

Ramsey, Paul. *The Patient as Person: Explorations in Medical Ethics*. New Haven, CT: Yale University Press, 1970.

Raphael, D. D. "Hume and Adam Smith on Justice and Utility." *Proceedings of the Aristotelian Society* 73 (1973): 87–103.

———, ed. *British Moralists, 1650–1800*. Oxford: Clarendon Press, 1969.

Read, Daniel. "Experienced Utility: Utility Theory from Jeremy Bentham to Daniel Kahneman." *Thinking & Reasoning* 13, no. 1 (2007): 45–61.

Reeves, Richard. *John Stuart Mill: Victorian Firebrand*. New York: Atlantic, 2008.

Reiser, Stanley Joel, Arthur J. Dyck, and William J. Curran, eds. *Ethics in Medicine: Historical Perspectives and Contemporary Concerns*. Cambridge, MA: MIT Press, 1977.

"Report from the Select Committee Appointed by the House of Commons to Inquire into the Manner of Obtaining Subjects for Dissection in the Schools of Anatomy, &c." In *Papers Relative to the Question of Providing Adequate Means for the Study of Anatomy*, 10–25. London: C. Adlard, 1832.

Reverby, Susan. *Examining Tuskegee: The Infamous Syphilis Study and Its Legacy*. Chapel Hill: University of North Carolina Press, 2009.

"Review of Report on the Poor Laws." *Christian Observer*,1818.

"Review of William Mackenzie 'an Appeal to the Public and to the Legislature, on the Necessity of Affording Dead Bodies to the Schools of Anatomy, by Legislative Enactment.'" *Literary Chronicle and Weekly Review*, no. 298 (January 29th, 1825).

Richards, Robert. *Darwin and the Emergence of Evolutionary Ideas of Mind and Behavior*. Chicago: University of Chicago Press, 1987.

Richardson, Benjamin Ward. *The Health of Nations: A Review of the Works of Edwin Chadwick, with a Biographical Dissertation*. London: Longmans, Green & Co., 1887.

Richardson, Ruth. *Death, Dissection, and the Destitute*. London: Routledge & Kegan Paul, 1987.

"Riot at Swan Street Cholera Hospital." *Manchester Guardian*, September 9, 1832.

Roberts, David. "How Cruel Was the Victorian Poor Law?" *Historical Journal* 6, no. 1 (1963): 97–107.

Robins, R. W., S. D. Gosling, and K. H. Craik. "An Empirical Analysis of Trends in Psychology." *American Psychology* 54, no. 2 (1999): 117–28.

Rodgers, Daniel T. *Contested Truths: Keywords in American Politics since Independence*. New York: Basic Books, 1987.

Romilly, Samuel. *Memoirs of the Life of Sir Samuel Romilly, Written by Himself*. London: John Murray, 1840.

Rosen, Frederick. "Jeremy Bentham on Slavery and the Slave Trade." In *Utilitarianism and Empire*, edited by Bart Schultz and Georgios Varouxakis, 33–56. Lanham, MD.: Lexington Books, 2005.

Rosenhan, D. L. "On Being Sane in Insane Places," *Science* 179, no 4070 (January 19, 1973): 250–58.

Rossinger, Leonard, and Myron Davies. "Prison Malaria: Convicts Expose Themselves to Disease So Doctors Can Study It." *Life* 18, no. 23 (June 4, 1945): 43–46.

Rothman, David J. *Strangers at the Bedside: A History of how Law and Bioethics Transformed Medical Decision Making*. New York: Basic Books, 1991.

Rothschild, Emma. *Economic Sentiments: Adam Smith, Condorcet and the Enlightenment*. Cambridge, MA: Harvard University Press, 2001.

Rushton, Bill. "The Mysterious Experiments of Dr. Heath, in Which We Wonder Who Is Crazy and Who Is Sane." *Courier*, August 29–September 4, 1974, 5–12.

Russell, Bertrand. *The Autobiography of Bertrand Russell*. Vol. 1, *1872–1914*. New York: Bantam Books, 1967.

———. *An Outline of Philosophy*. London: Routledge, 1993.

Saluka, Alex. "Baroness Burdett-Coutt's Garden Party: The International Medical Congress, London, 1881." *Medical History* 26, no. 2 (1982): 183–90.

Sanders, David, and Jesse Durkeminier. "Medical Advance and Legal Lag: Hemodialysis and Kidney Transplantation." *UCLA Law Review* 15, no. 2 (1968): 357–413.

Sass, Hans-Martin. "Reichsrundschreiben 1931: Pre-Nuremberg German Regulations Concerning New Therapy and Human Experimentation." *Journal of Medicine and Philosophy* 8, no. 2 (1983): 99–112.

Schaffer, Simon. "States of Mind: Enlightenment and Natural Philosophy." In *The

Languages of Psyche: Mind and Body in Enlightenment Thought: Clark Library Lectures, 1985–1986, edited by George Sebastian Rousseau, 233–90. Berkeley: University of California Press, 1990.

Schmidt, Ulf. *Justice at Nuremberg: Leo Alexander and the Nazi Doctors' Trial.* New York: Palgrave Macmillan, 2004.

Schofield, Philip. *Utility and Democracy: The Political Thought of Jeremy Bentham.* Oxford: Oxford University Press, 2006.

Schultz, Bart,and Georgios Varouxakis. *Utilitarianism and Empire.* Lanham, MD.: Lexington Books, 2005.

Scull, Andrew T. *The Most Solitary of Afflictions: Madness and Society in Britain, 1700–1900.* New Haven, CT: Yale University Press, 1993.

Scull, Andrew T., Charlotte MacKenzie, and Nicholas Hervey. *Masters of Bedlam: The Transformation of the Mad-Doctoring Trade.* Princeton, NJ: Princeton University Press, 1996.

Selby-Bigge, L. A. *British Moralists: Being Selections from Writers Principally of the Eighteenth Century.* Oxford: Clarendon Press, 1897.

Semple, Janet. *Bentham's Prison: A Study of the Panopticon Penitentiary.* Oxford: Clarendon Press, 1993.

Sensen, Oliver. *Kant on Moral Autonomy.* Cambridge: Cambridge University Press, 2013.

Shapin, Steven. *A Social History of Truth: Civility and Science in Seventeenth-Century England.* Chicago: University of Chicago Press, 1994.

Shapin, Steven, and Simon Schaffer. *Leviathan and the Air-Pump.* Princeton, NJ: Princeton University Press, 1985.

Shoda, Yiuchi, Walter Mischel, and Philip K. Peake. "Predicting Adolescent Cognitive and Self-Regulatory Competencies from Preschool Delay of Gratification: Identifying Diagnostic Conditions." *Developmental Psychology* 26, no. 6 (1990): 978–86.

Shweder, Richard A. "Tuskegee Re-examined." *Spiked,* January 8, 2004, http://www.spiked-online.com/newsite/article/14972#.VzWcU4QrLIV.

Sjöström, Henning and Robert Nilsson. *Thalidomide and the Power of the Drug Companies.* Harmondsworth: Penguin, 1972.

Skinner, B. F. *The Behavior of Organisms: An Experimental Analysis.* New York: Appleton-Century-Crofts, 1938.

———. *Beyond Freedom and Dignity.* London: Penguin, 1988.

———. *A Matter of Consequences.* New York: Alfred Knopf, 1983.

———. *The Shaping of a Behaviorist.* New York: Alfred A. Knopf, 1979.

———. *Walden Two.* New York: Macmillan, 1970.

Skinner, Quentin. *Visions of Politics: Hobbes and Civil Science.* Cambridge: Cambridge University Press, 2002.

Skloot, Rebecca. *The Immortal Life of Henrietta Lacks.* New York: Crown Publishers, 2010.

Smith, Adam. *The Theory of Moral Sentiments.* Edited by D. D. Raphael and A. L. Macfie. Oxford: Clarendon Press, 1979.

———. *The Wealth of Nations*. Books 1–3. London: Penguin, 1986.

Smith, D. W. "Helvétius and the Problems of Utilitarianism." *Utilitas* 5, no. 2 (1993): 275–89.

Smith, Thomas Southwood. *Use of the Dead to the Living: From the Westminster Review*. Albany: Webster and Skinners, 1827.

Spencer, Herbert. *Autobiography*. London: Williams and Norgate, 1904.

———. "Bain on the Emotions and the Will." *Medico-Chirurgical Review*, January 1860.

———. *The Data of Ethics*. London: G. Norman and Son, 1879.

———. *Principles of Psychology*. London: Williams and Norgate, 1870.

———. *Social Statics; or, The Conditions Essential to Human Happiness Specified, and the First of Them Developed*. London: John Chapman, 1851.

———. "A Theory of Population, Deduced from the General Law of Animal Fertility." *Westminster Review* 57, no. 112 (1852): 250–68.

Spillane, John D. *The Doctrine of the Nerves*. Oxford: Oxford University Press, 1981.

Spitzer, Robert L. "The Diagnostic Status of Homosexuality in DSM-III: A Reformulation of the Issues." *American Journal of Psychiatry* 138, no. 2 (1981): 210–15.

Star, Susan Leigh. *Regions of the Mind: Brain Research and the Quest for Scientific Certainty*. Stanford, CA: Stanford University Press, 1989.

Stark, Laura. *Behind Closed Doors: IRBs and the Making of Ethical Research*. Chicago: University of Chicago Press, 2012.

Stephen, Leslie. *The English Utilitarians*. London: Duckworth and Co., 1900.

Stokes, Eric. *The English Utilitarians and India*. Oxford: Clarendon Press, 1963.

Szasz, Thomas. "Against Behaviorism: A Review of B. F. Skinner's *About Behaviorism*." *Libertarian Review* 111 (December 1974).

Taylor, Barbara. *The Last Asylum: A Memoir of Madness in Our Times*. Chicago: University of Chicago Press, 2015.

Taylor, Telford. *The Anatomy of the Nuremberg Trials: A Personal Memoir*. New York: Knopf, 1992.

Thaler, Richard H. *Misbehaving: The Making of Behavioral Economics*. New York: W. W. Norton, 2015.

Thaler, Richard H., and H. M. Shefrin. "An Economic Theory of Self-Control." *Journal of Political Economy* 89, no. 2 (April 1981): 392–406.

Thorndike, Edward Lee. *Animal Intelligence: An Experimental Study of the Associative Processes in Animals*. New York: Macmillan, 1898.

———. *Animal Intelligence: Experimental Studies*. New York: Macmillan, 1911.

———. "Science and Values." *Science* 83, no. 2140 (1936): 1–8.

Thorwald, Jürgen, and Leon Banov. *The Triumph of Surgery*. New York: Pantheon, 1960.

Tomlinson, Wallace. "Interview with Robert Galbraith Heath." Video. 1986.

Townsend, Joseph. *A Dissertation on the Poor Laws by a Well-Wisher to Mankind*. Berkeley: University of California Press, 1971.

Transactions of the International Medical Conference Seventh Session. London: Ballantyne Hanson, 1881.

Trials of War Criminals before the Nuernberg Military Tribunals under Control Council Law no. 10, Nuremberg, October 1946–April 1949. Washington, DC: US Government Printing Office, 1949.

Tuke, Samuel. "Plans, Elevations, Sections and Descriptions of the Pauper Lunatic Asylum Lately Erected at Wakefield by Messrs. Watson and Pritchett, Including Instructions Drawn Up by Mr. Samuel Tuke for the Architects Who Prepared Designs for the West Riding Asylum" (1819). Yorkshire Archive Service, Wakefield.

———. *Practical Hints on the Construction and Economy of Pauper Lunatic Asylums, including Instructions to the Architects who offered Plans for the Wakefield Asylum.* York: William Alexander, 1815.

Tully, James. "Placing the Two Treatises." In *Political Discourse in Early Modern Britain,* edited by Nicholas Phillipson and Quentin Skinner, 253–80. Cambridge: Cambridge University Press, 1993.

Tyler, Kenneth, and Rolf Malessa. "The Goltz-Ferrier Debates and the Triumph of Cerebral Localizationist Theory." *Neurology* 55, no. 7 (2000): 1015–24.

Vollmann, Jochen, and Rolf Winau. "Informed Consent in Human Experimentation before the Nuremberg Code." *BMJ* 313, no. 7070 (1996): 1445–47.

Vorzimmer, Peter. "Darwin, Malthus, and the Theory of Natural Selection." *Journal of the History of Ideas* 30, no. 4 (October–December 1969): 527–42.

Wailoo, Keith. *Pain, a Political History.* Baltimore: Johns Hopkins University Press, 2014.

Watson, John B. *Behavior: An Introduction to Comparative Psychology.* New York: H. Holt and Co., 1914.

———. "Psychology as the Behaviorist Views It." *Psychological Review* 20, no. 2 (1913): 158–77.

Wedgwood, Thomas. *The Value of a Maimed Life: Extracts from the Manuscript Notes of Thomas Wedgwood.* Edited by Margaret Olivia Tremayne. London: C. W. Daniel, 1912.

Weindling, Paul. "The Origins of Informed Consent: The International Scientific Commission on Medical War Crimes and the Nuremberg Code." *Bulletin of the History of Medicine* 75, no. 1 (2001): 37–71.

Weinstein, David. *Equal Freedom and Utility: Herbert Spencer's Liberal Utilitarianism.* Cambridge: Cambridge University Press, 1998.

Wilson, Catherine. *Epicureanism at the Origins of Modernity.* Oxford: Oxford University Press, 2008.

Woodworth, R. "Edward Lee Thorndike, 1874–1949." *Biographical Memoirs of the National Academy of Sciences of the United States of America* 27 (1952): 209–37.

Wooten, James T. "Survivor of '32 Syphilis Study Recalls a Diagnosis." *New York Times,* July 27, 1972.

Yonometo, Isaac T., et al. "Dual Organism Design Cycle Reveals Small Subunit Substitutions That Improve [NiFe] Hydrogenase Hydrogen Evolution." *Journal of Biological Engineering* 7, no. 17 (2013).

Pages in italics refer to captions.